# CONTENTS

Introduction ............................................................................................. 13
Supplies list ............................................................................................. 15

**Activity 1**
Bouncing off the Walls and Everything! .............................................. 19

**Activity 2**
Light to Warn and Guide ...................................................................... 22

**Activity 3**
Color ........................................................................................................ 24

**Activity 4**
Shadows .................................................................................................. 28

**Activity 5**
Reflection (Bouncing Light) ................................................................. 30

**Activity 6**
Bending Light ......................................................................................... 32

**Activity 7**
Day and Night ........................................................................................ 37

**Activity 8**
Fuel and Weird Light ............................................................................ 39

GOD MADE EVERYTHING

**Activity 9**
Science Tools ............................................................................................... 43

**Activity 10**
Seeing Created Beauty ............................................................................... 46

**Activity 11**
Take Care of Your Eyes! ............................................................................. 50

**Activity 12**
Useful Animal Eyes ..................................................................................... 52

**Activity 13**
The Big Earth ............................................................................................... 55

**Activity 14**
The Exploding Earth ................................................................................... 58

**Activity 15**
The Pulling Earth ........................................................................................ 60

**Activity 16**
Rock Hunt .................................................................................................... 63

**Activity 17**
Mine for Metal ............................................................................................ 66

**Activity 18**
Make a Flood and Fossil Frittata .............................................................. 68

**Activity 19**
Peace; Be Still! ............................................................................................. 72

**Activity 20**
Water for Life .............................................................................................. 75

**Activity 21**
Cycle Some Water..................................................................................78

**Activity 22**
Happy Birthday! (to Someone)..............................................................80

**Activity 23**
Seasons of the Year................................................................................82

**Activity 24**
Hibernate!................................................................................................85

**Activity 25**
Your Five Senses.....................................................................................86

**Activity 26**
How to Put Out a Fire............................................................................88

**Activity 27**
Wing It!....................................................................................................90

**Activity 28**
Five Layers...............................................................................................91

**Activity 29**
Cotton Clouds.........................................................................................94

**Activity 30**
Snowflake Art Project............................................................................96

**Activity 31**
Man on the Moon...................................................................................98

**Activity 32**
Eclipse of the Sun.................................................................................101

# GOD MADE EVERYTHING

**Activity 33**
Moon Shapes ..................................................................................................103

**Activity 34**
The Lonely Little Space Rock ......................................................................105

**Activity 35**
Jump to the Planets! .....................................................................................109

**Activity 36**
Draw a Star ....................................................................................................111

**Activity 37**
Roots Store Food ...........................................................................................113

**Activity 38**
Watch a Plant Drink .....................................................................................116

**Activity 39**
Gardening .......................................................................................................118

**Activity 40**
Plant Some Seeds ..........................................................................................120

**Activity 41**
Stretch Like a Stem .......................................................................................122

**Activity 42**
Let's Play Pollination! ...................................................................................125

**Activity 43**
Greenery .........................................................................................................131

**Activity 44**
Trading Gasses ...............................................................................................135

**Activity 45**
Colored Leaves..................................................................................................137

**Activity 46**
Rhyming Practice ............................................................................................141

**Activity 47**
Ah-Choo!..........................................................................................................143

**Activity 48**
God Gives Us Other Plant Gifts ....................................................................145

**Activity 49**
Examine Some Exoskeletons .........................................................................147

**Activity 50**
Naming ............................................................................................................151

**Activity 51**
Wise Creatures ................................................................................................155

**Activity 52**
How Many? .....................................................................................................157

**Activity 53**
Big Changes!....................................................................................................159

**Activity 54**
Be a Mess-Eating Insect!.................................................................................163

**Activity 55**
Lots of Little Locusts .....................................................................................165

**Activity 56**
Bee Maze..........................................................................................................167

**Activity 57**
What Am I? .................................................................................................. 169

**Activity 58**
Air in Water .................................................................................................. 171

**Activity 59**
Make a Web! ................................................................................................ 173

**Activity 60**
Butterfly Scales ............................................................................................ 175

**Activity 61**
Swim Bladders ............................................................................................. 179

**Activity 62**
You Have Something Sharks Have! ............................................................. 182

**Activity 63**
Precious Pearls ............................................................................................. 184

**Activity 64**
Grow into a Frog .......................................................................................... 188

**Activity 65**
Webbed Feet ................................................................................................ 190

**Activity 66**
Salamander Coloring Page ........................................................................... 192

**Activity 67**
Find the Warmth .......................................................................................... 195

**Activity 68**
Sense Like a Snake ...................................................................................... 197

**Activity 69**
Tortoise or Turtle? ...................................................................................................199

**Activity 70**
Make Print Fossils .....................................................................................................201

**Activity 71**
Behemoth Sculpture..................................................................................................203

**Activity 72**
Draw Leviathan ........................................................................................................206

**Activity 73**
Wings Push Air..........................................................................................................208

**Activity 74**
Learn to Whistle!......................................................................................................212

**Activity 75**
Bird Food...................................................................................................................214

**Activity 76**
Beaks Are Tools ........................................................................................................216

**Activity 77**
Bird Feet....................................................................................................................218

**Activity 78**
Eggs ...........................................................................................................................221

**Activity 79**
Help Migrating Birds ...............................................................................................224

**Activity 80**
Finding North ...........................................................................................................227

GOD MADE EVERYTHING

**Activity 81**
Nesting ...................................................................................228

**Activity 82**
Oil and Water Don't Mix ......................................................231

**Activity 83**
Birds Named for Their Songs ..............................................233

**Activity 84**
Be a Honeyguide! .................................................................234

**Activity 85**
Mammals Feed Their Babies ...............................................237

**Activity 86**
Quills for Protection ............................................................241

**Activity 87**
Animal Baby Names ............................................................243

**Activity 88**
Special Heads and Feet ........................................................245

**Activity 89**
Thirsty Camels .....................................................................248

**Activity 90**
Color the Wild Hoofed Mammals .......................................253

**Activity 91**
Paw Pads ...............................................................................255

**Activity 92**
Hyenas Laugh and Eat .........................................................257

**Activity 93**
Dry Some Fruit!......................................................................................................261

**Activity 94**
Do the Elephant Walk and the Monkey Swing!....................................................264

**Activity 95**
Sea Mammals.........................................................................................................266

**Activity 96**
Hang Like a Bat!....................................................................................................269

**Activity 97**
Made in His Image................................................................................................271

**Activity 98**
There's No One Else Like You!..............................................................................273

**Activity 99**
You Were Once a Newborn...................................................................................277

**Activity 100**
Neuron Chain Tag.................................................................................................281

**Activity 101**
Sound Tag..............................................................................................................283

**Activity 102**
The Tongue's Jobs..................................................................................................285

**Activity 103**
Growing.................................................................................................................287

**Activity 104**
Take a Breath.........................................................................................................289

**Activity 105**
Listen to Your Stomach ..................................................................................................291

**Activity 106**
Your Skeleton ..................................................................................................................293

**Activity 107**
Using Muscles ................................................................................................................295

**Activity 108**
Trying to Learn ...............................................................................................................297

Answer Key .....................................................................................................................301

# INTRODUCTION

This introductory science course for young school-age children is designed to bring to light the love, wisdom, and power of God that is evident in His creation. *God Made Everything* presents the amazing way each created thing works perfectly and how all the created things work harmoniously together in a way that only God could have brought forth.

Designed to complement the way God created children to learn, this activity book provides hands-on learning to go along with the auditory presentation and beautiful visuals in the read-aloud textbook.

Beamer, the friendly light beam, visits this activity book as well as the textbook. His love for God's creation shines here too!

The *God Made Everything Activity Book* features 108 activities that involve the senses and muscles yet are designed to be not too burdensome for the parent/teacher. Each branch of science is reinforced with a balance of action, observation, experiments, imagination, logic, Scripture, art, cooking, poetry, math, stories, exercise, music, and a little bit of writing. Each activity reinforces the material introduced in the textbook, and every exercise is numbered for easy reference.

## How to Use the God Made Everything Activity Book

*God Made Everything* is divided into nine units of four chapters each. Each unit has a memory verse and a chil-

dren's hymn that children will have an opportunity to work on while completing the activities.

The *God Made Everything* books are organized in a way to enable children to internalize what they have learned. Learning comes easiest in small doses with time between each session. Children solidify what they have learned as they play. It becomes permanent as they sleep. To enable this structure of learning, the following schedule is suggested below.

For convenience, this activity book includes a supplies list and answer key.

May the Lord be glorified, and may you be richly blessed as you and your little learners study God's creation in *God Made Everything*!

## Suggested Schedule

One chapter per week for 36 weeks.

| | |
|---|---|
| Day 1 | Read aloud the first section of the textbook chapter. Complete the corresponding activity in the activity book (as announced by number at the end of that section of text). |
| Day 2 | Break |
| Day 3 | Read aloud the second section of the textbook chapter. Complete the corresponding activity in the activity book. |
| Day 4 | Break |
| Day 5 | Read aloud the third section of the textbook chapter. Complete the corresponding activity in the activity book. |

# SUPPLIES LIST

## SUPPLIES TO HAVE ON HAND THROUGHOUT THE ACTIVITY BOOK:

- Bible
- Pencil
- Blank paper for artwork, notebook size
- Colored pencils, full selection of colors
- Crayons, full selection of colors
- Pair of scissors
- Glue
- Tape
- Water

## CHAPTER 1

- A ball and a wall to roll it against
- An outdoor place with enough objects to bump into
- Flashlight and a dark room or outdoor place
- White card stock for a 6" diameter circle

## CHAPTER 2

- A bright light to shine on a wall
- Two large pieces of white paper, at least 11"x14"
- A toy suitable for making a traceable shadow
- Small mirror
- Access to a large wall mirror
- Flashlight and a dark room
- Access to a garden hose and spray nozzle, or a spray bottle

## CHAPTER 3

- Styrofoam ball, diameter 3"-6" or a round fruit like an orange
- Skewer
- Pushpin
- Flashlight

## CHAPTER 4

- Shovel or spade
- An outdoor place to dig
- An outdoor place to observe nature and beauty
- Clipboard
- Drawing paper, pencil, crayons
- Camera or phone that can take pictures
- Garden clipper
- Flat box or empty egg carton
- Small ball
- Small mirror

15

# GOD MADE EVERYTHING

## CHAPTER 5
- Waterproof marker
- A small bottle or narrow jar that can hold 1/4 cup (2 oz. or 60 ml)
- 1/2 teaspoon baking soda (2-1/2 grams)
- Red and yellow food coloring
- 1/4 cup (2 oz. or 60 ml) vinegar
- Water and dirt
- A place to make mud

## CHAPTER 6
- Empty egg carton
- Tray to set egg carton on
- 12 dried beans or a package of bean seeds
- Small magnet
- Small plastic bag that the magnet will fit into
- Access to some dry dirt
- A 9x13x3 baking pan
- 1 hot dog bun
- 1/4 cup (60 g) butter
- A small handful of tiny seeds (cumin, poppy, quinoa, chia)
- 16 oz (450 g) grated cheddar cheese
- 1 10.5 oz (298 g) can of cream of chicken soup
- 3 oz (80g) tortilla chips
- 1/4 cup (40 g) corn meal
- 8 oz (230 g) can of chopped green chiles
- 6 eggs, beaten
- 1 cup (0.25 L) milk
- 1/2 teaspoon (3 g) salt
- Lettuce, finely shredded
- Diced, fresh tomatoes

## CHAPTER 7
- A cup, pan, or bucket
- Small mirror
- Spoon
- Facial tissue
- Refrigerator
- Small self-sealing plastic bag
- Blue food coloring
- Empty drinking glass
- Sunny window

## CHAPTER 8
- Flashlight
- Blankets, chairs, boxes to build a hibernation den
- A snack
- Pillow

## CHAPTER 9
- Votive candle or tea light
- Metal pan
- Glass jar
- Cardboard toilet paper or paper towel tube
- Stapler

## CHAPTER 10
- Glass spice jar or small narrow jar
- Funnel
- 3 Tbsp. (24g) ground cinnamon
- 2 tsp. (4g) sugar
- 2 tsp. (4g) ground ginger
- 2 tsp. (5g) nutmeg
- 1-1/2 tsp. (3g) ground cloves
- 1-1/2 tsp. (3g) ground allspice
- An outdoor swing
- A sheet of blue paper
- About 6 cotton balls

## CHAPTER 11
- Flashlight
- 2-3 facial tissues
- An opaque marble
- A ball (not shiny)

# SUPPLIES LIST

## CHAPTER 12
- A place to take about 10 jumps forward
- Sidewalk chalk or pieces of string to mark 8 increments
- A die to roll

## CHAPTER 13
- A shallow dish
- 2-3 root vegetables (carrot, beet, turnip, etc.)
- Sharp knife
- Bright window
- Clear drinking glass or jar
- Stalk of celery
- Blue or red food coloring

## CHAPTER 14
- Pots, egg carton, and plastic containers to plant seeds in
- Potting soil or dirt
- Seeds—purchased or saved from foods
- Four envelopes
- Two glasses of water or juice

## CHAPTER 15
- Five cotton swabs
- Selection of fresh green leaves

## CHAPTER 16
- Nothing extra

## CHAPTER 17
- Two or three small plastic containers with lids
- Cotton swab
- A piece of stiff paper
- Magnifying glass
- A stiff comb

## CHAPTER 18
- A messy room to clean up

## CHAPTER 19
- Nothing extra

## CHAPTER 20
- Large frying pan
- Stove
- Three hard-backed chairs with posts
- Spool of thread
- Sequins

## CHAPTER 21
- Two uninflated balloons
- Two of the same kind of coin
- Waterproof marker
- Bucket

## CHAPTER 22
- Large adult t-shirt
- Pillowcase
- Sturdy rubber band
- Plastic bag to fit child's spread hand

## CHAPTER 23
- Nothing extra

## CHAPTER 24
- A sheet of aluminum foil
- Purchased modeling dough, or
- Homemade dough made with:
- 1 c. (125g) flour
- 1/4 c. (75g) salt
- 3T. (45ml) lemon juice
- 1T. (15ml) cooking oil
- Water based paint (optional)

## CHAPTER 25
- Nothing extra

## CHAPTER 26
- Popped popcorn
- Small bowl
- Cardboard egg carton with tall pointed dividers
- A sturdy pair of scissors
- Broom stick
- Two stacks of books 6" tall
- An egg
- 3/4 cup (170 g) butter, at room temperature
- 1/4 cup (50 g) sugar
- 1/2 teaspoon (2.5 ml) pure vanilla extract
- 1 3/4 cups (240 g) all-purpose flour
- 1/2 teaspoon (3 g) salt
- 1 egg beaten
- 1/2 cup (50 g) shredded coconut
- 5 ounces (150 g) jelly beans
- Cookie making supplies

## CHAPTER 27
- Compass for finding north
- Sidewalk chalk or a stick for marking on the ground

## CHAPTER 28
- Vegetable oil
- A bottle of honey

## CHAPTER 29
- A ripe pear or kiwi
- A box of toothpicks
- Eleven raisins
- A plate

## CHAPTER 30
- A one gallon (4L) container

## CHAPTER 31
- Baking sheet
- Baking parchment
- Sharp knife
- Oven
- Fresh fruit like berries, cherries, apples, peaches, apricots, pears, or bananas

## CHAPTER 32
- Access to a recording of "Baby Elephant Walk" by Henry Mancini
- Access to a playground bar or tree branch the child can grab

## CHAPTER 33
- Nature pictures from magazines or the internet
- Ink pad for making fingerprints

## CHAPTER 34
- Large area for a game of tag
- Smaller area for tag. It would need boundaries so you can say, "Stay on the grass" or "Stay on the basketball court."
- A snack

## CHAPTER 35
- Measuring tape
- Bathroom scale
- Glass of water
- Roll of paper towels at least half full

## CHAPTER 36
- Yardstick or broomstick
- String
- 10-20 white pipe cleaners

UNIT 1 • CHAPTER 1
ACTIVITY 1
# Bouncing off the Walls and Everything!

Light bounces off things the same way a ball does. If you roll a ball straight at a wall, it comes straight back to you. If you roll it a little sideways (we say at an angle), it will leave the wall at an angle. If you roll a ball at a bumpy rock, it will not come straight back because of the angles of the rock.

 Try rolling a ball against a wall at different angles. Try rolling it against objects with different shapes.

Now let's pretend to be light. Remember, light usually travels in a straight line. When it hits something, it bounces off and keeps traveling in a straight line until it hits something else. This happens many times and is the way light spreads and wins over the dark.

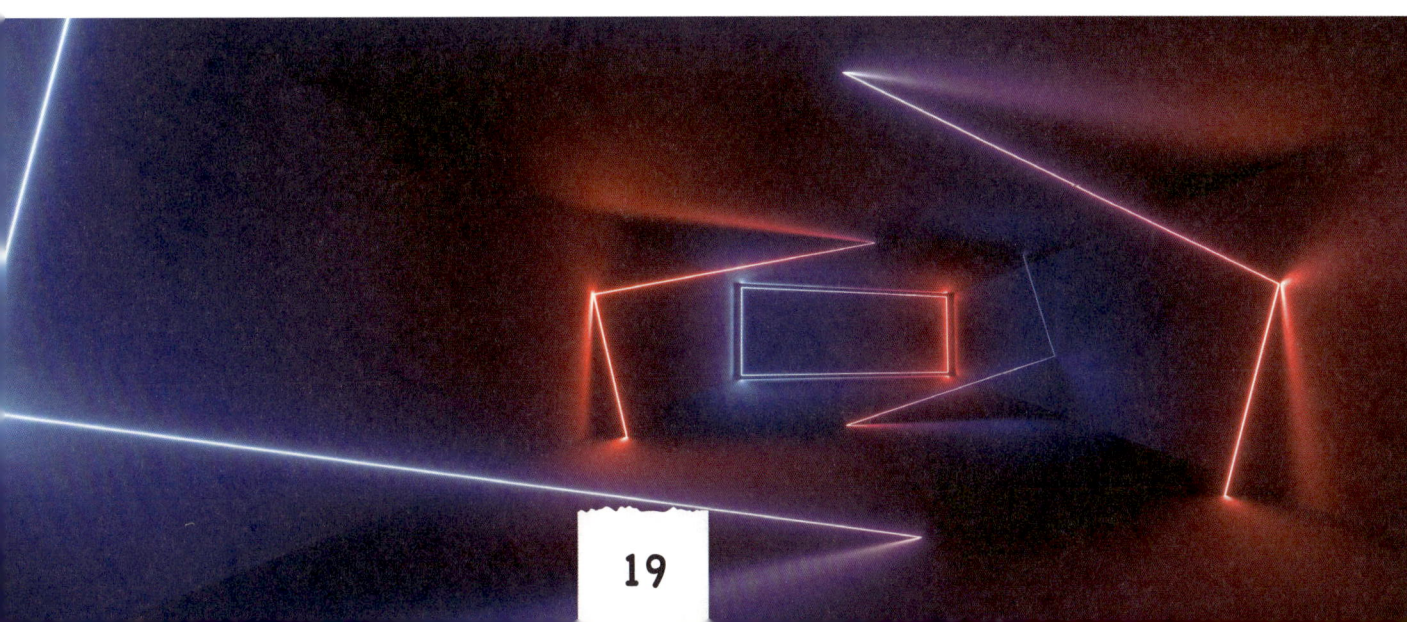

# GOD MADE EVERYTHING

 Pretend to be light. Walk around the room, slowly bouncing off walls and furniture. Think about the angles light would bounce off things. (It bounces the same as a ball.)

 Since light is fast, go outside where you can run. Pretend to be light. Be careful when you bounce off things!

 Sing the hymn for this unit.

 Write or repeat the memory verse for this unit.

### Memory Verse

Arise, shine,
for your light has come!
And the glory of the Lord
is risen upon you.

(Isaiah 60:1)

## ACTIVITY 1: BOUNCING OFF THE WALLS AND EVERYTHING!

Arise, shine,
for your light has come!
And the glory of the Lord
is risen upon you.
Isaiah 60:1

Thank God for His goodness in making beautiful light!

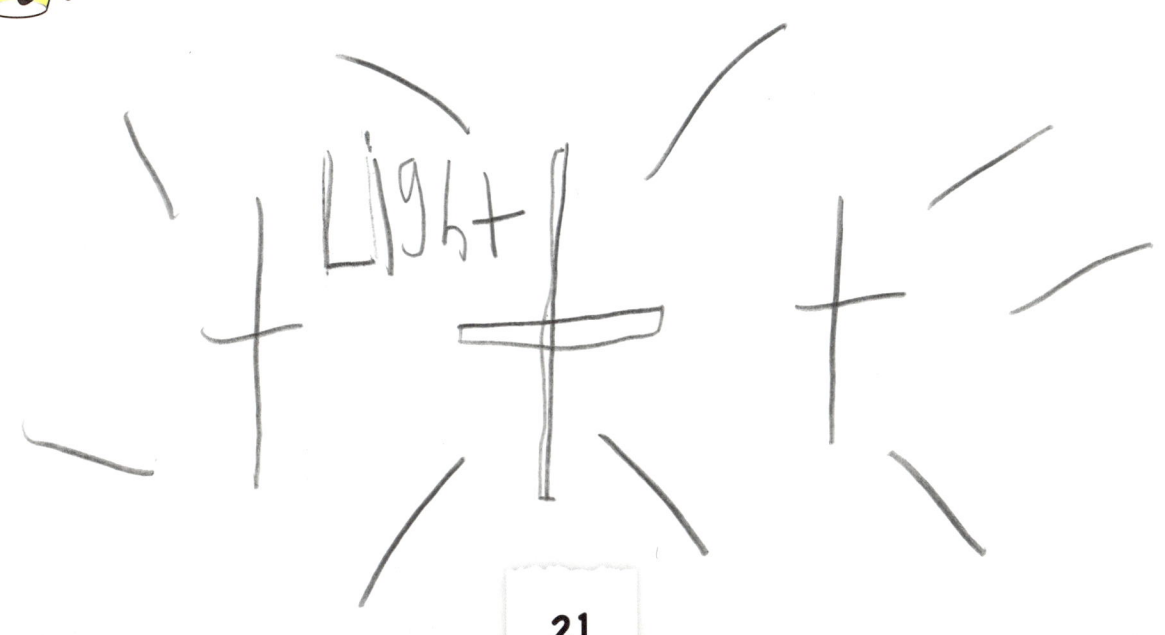

21

UNIT 1 • CHAPTER 1

# ACTIVITY 2
# Light to Warn and Guide

**Your word is a lamp to my feet and a light to my path. (Psalm 119:105)**

 Go to a dark room or a dark outdoor place with a flashlight. Turn on the flashlight so you can see where to walk. Now turn it off and see how hard it would be to walk safely in the dark. God's Word is a light to us. It warns us of dangers and shows us how to live.

 Five boats were lost on the sea. Three of them saw the light beam from a lighthouse and have gone to safety. How many of them have not seen the light beam yet? _____ Draw them in the picture below.

### ACTIVITY 2: LIGHT TO WARN AND GUIDE

 Sing the hymn for this unit.

 Repeat the memory verse for this unit.

 Thank God for giving us His Word to lead and guide us. Thank Him for the safety light gives.

UNIT 1 • CHAPTER 1

ACTIVITY 3
# Color

 Even though light looks white to us, it has every color inside it. Let's make a color spinner to show this.

### Color Spinner

1. Tear out the page with the colored circle.
2. With crayons, finish coloring each pie-shaped piece in the same color.
3. Cut out your colored circle.
4. Cut a circle of the same size out of a piece of white card stock.
5. Glue your colored circle onto the card stock.
6. With help, push a pencil through the center of the circle so the point is on the bottom.
7. Take the spinner outside where God's light gives us all the colors. A shady spot works best. Spin the pencil as fast as you can between your hands or like a top. An adult may have to spin it faster than you can so that you don't see the separate colors. You may need to apply glue where the pencil pokes through the circle to keep it from slipping when you spin it.

1. What color is the circle when it is spinning fast?

   _____

ACTIVITY 3: COLOR

2. Why do you think this is?

_____

(Don't worry if it looks a little brown. Your colors may not be pure.)

3. What is your favorite color? Write or color it here:

_____

 Sing the hymn for this unit.

 Repeat the memory verse for this unit.

 Thank God for beautiful colors. Thank Him for something that is blue. Find something for each color on your spinner and thank Him for all these beautiful things.

UNIT 1 • CHAPTER 2

# ACTIVITY 4
# Shadows

 Why doesn't Beamer see his shadow?

_____

 On a sunny day, go outside in the middle of the day. Look at your shadow. Where is it? Go back outside late in the day. Look at your shadow again. Is it in the same place as it was in the middle of the day? Circle the picture below that was taken around lunch time.

ACTIVITY 4: SHADOWS

 Let's make a silhouette (an outline filled in with black or another color). You will need a bright light and a large toy.

### Make a Silhouette

1. Set a bright light on a table so it can shine on a wall.
2. Tape a large piece of paper to the wall on a level with the light.
3. Put a large toy on the table in front of the light so that it casts a shadow on the paper. You may need to move the toy around until you get a good edge to the shadow.
4. Use a pencil to trace the toy's shadow on your paper.
5. Take down the paper and color the inside of your outline.
6. You may want to ask someone to trace the shadow of your head's side view on another piece of paper so you have a silhouette of yourself at this age.

 Sing the hymn for this unit.

 Repeat the memory verse for this unit.

 Thank God for cool shade in summer. Thank Him for His shadow of protection.

UNIT 1 • CHAPTER 2

# ACTIVITY 5
# Reflection (Bouncing Light)

 Get a small mirror and experiment with it. Use the mirror to see behind you. Use it to see under a table. Ask someone to stand around a corner and make a silly face. Can you see their silly face using only your mirror?

 Stand in front of a large mirror. Hold your small mirror against your nose, just below your eyes so that it is facing the large mirror. Look at yourself in the large mirror. Tilt the small mirror different ways until you see a tunnel of small mirrors and many eyes. The light is bouncing back and forth between the mirrors! It seems that the tunnel of mirrors could go on forever. This reminds us of another type of forever. God gives His people eternal life—life that goes on forever!

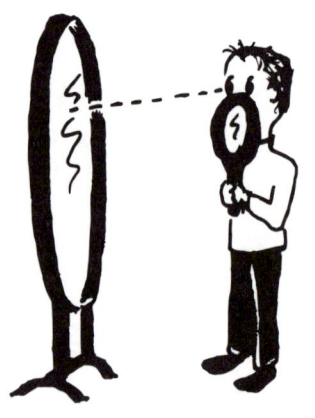

 Take a flashlight and a small mirror into a dark room. Put the mirror on the floor. Shine your flashlight on the mirror from different parts of the room. Look for the bright spot where the light hits the walls and ceiling.

## ACTIVITY 5: REFLECTION (BOUNCING LIGHT)

 Color the rainbow coloring page.

 Repeat the memory verse for this unit.

 Thank God for beautiful reflections on water and for the usefulness of mirrors. Thank Him for eternal life.

UNIT 1 • CHAPTER 2

# ACTIVITY 6
# Bending Light

> Like the appearance of a rainbow in a cloud on a rainy day, so was the appearance of the brightness all around it. This was the appearance of the likeness of the glory of the LORD. (Ezekiel 1:28)

 Let's make a real rainbow! You will need a hose with a spray nozzle or a spray bottle.

1. Watch for a day when it is sunny in the middle of the morning or in the middle of the afternoon.

2. Put a spray nozzle on your hose or fill a spray bottle with water.

3. Stand with the sun behind you. Your shadow should be in front of you.

4. Spray water in front of you and look for a rainbow! Notice that the shadow of your head is right in the middle of the rainbow.

ACTIVITY 6: BENDING LIGHT

 Did you know that rainbow colors are always in the same order? Color the rainbow coloring page. Use the list of colors to choose from. The first letter of each color is on the correct band of the rainbow.

 Sing the hymn for this unit.

 Repeat the memory verse for this unit.

 Thank God for giving us beautiful rainbows to show His promise.

33

# GOD MADE EVERYTHING

ACTIVITY 7: DAY AND NIGHT

Orange
Green
Violet
Indigo
Yellow
Red
Blue

35

# GOD MADE EVERYTHING

UNIT 1 • CHAPTER 3

ACTIVITY 7
# Day and Night

 Let's make a model of the earth. You will need a Styrofoam ball or a round fruit, and a skewer.

1. With help from an adult, mark the north and south poles on a Styrofoam ball or on a round fruit like an orange.

2. With help from the adult, poke a skewer through the Styrofoam ball, from pole to pole, as in the picture.

3. Press a pushpin into the side of your Earth model to show where you live on the real Earth.

4. Take your Earth model and a flashlight into a dark room.

5. Turn on the flashlight and shine its light onto the side of your Earth model. Then slowly spin your Earth from left to right. Earth spins at a tilt, so be sure to tilt your Earth a little as you spin.

Model of tilted Earth

GOD MADE EVERYTHING

6. Now ask someone to spin your Earth model for you. Watch your thumbtack. When the thumbtack first comes into the flashlight's beam, stretch your arms to show that it is sunrise for your thumbtack. When your thumbtack is in the middle of the flashlight's beam, pretend to eat lunch. Yawn when it is sunset for your thumbtack.

7. Save your Earth model for a future chapter when we learn about seasons.

 Sing the hymn for this unit.

 Repeat the memory verse for this unit.

 Praise God for the earth spinning in and out of sunlight. Thank Him for the daytime when you can play, learn, and work. Thank Him for the night when you can sleep. Praise Him for His care of us!

38

UNIT 1 • CHAPTER 3
ACTIVITY 8
# Fuel and Weird Light

Most fuel started as plants. Before Noah's flood, large forests grew on the earth. They were torn up during the flood and then buried with sand. They became coal. Some of the coal was buried deeper and deeper. Its temperature became so hot that oil and gas were formed.[1] Today we burn oil, gas, and coal as fuel.

Answer these questions:

1. Coal, gas, and oil can burn because they started as

   _____.

2. Peat moss is from plants that rotted underwater in swamps. People dig it up and dry it. They use it as fuel for their fires. Peat moss started as

   _____.

---

1. "The Origin of Oil," *Answers* 2:1, 2007, pp. 74–77, online at www.answersingenesis.org/geology/the-origin-of-oil/.

# GOD MADE EVERYTHING

3. Cows eat plants. After their bodies get everything they need from the plants, the waste comes out as manure. People can use manure for fuel. Manure started as

   _____ .

 Color the anglerfish coloring page.

 Sing the hymn for this unit.

 Repeat the memory verse for this unit.

 Thank God for making Earth the right distance from the sun so that we (and all life) can be here. Thank God for the warmth He provides for our life and comfort. Light from living creatures is amazing! Praise Him for His wisdom in making interesting creatures that glow without burning up.

ACTIVITY 8: FUEL AND WEIRD LIGHT

# GOD MADE EVERYTHING

UNIT 1 • CHAPTER 3
ACTIVITY 9
# Science Tools

### Definition

**Science** is learning (or studying) God's creation.

**Observing** is carefully looking, listening, feeling, smelling, and sometimes tasting something. This helps us learn about it.

**Measuring** is using tools to find a number for distance, size, weight, amount, temperature, brightness, or loudness.

*The works of the LORD are great, studied by all who have pleasure in them. (Psalm 111:2)*

Look at the pictures of the science tools. Write an "O" under the tool if it is used for observing. Write an "M" if it is used for measuring.

# GOD MADE EVERYTHING

1. _____

2. _____

3. _____

4. _____

5. _____

6. _____

6. _____

44

ACTIVITY 9: SCIENCE TOOLS

Sing the hymn for this unit.

Repeat the memory verse for this unit.

Thank God for His wisdom in creating everything! Thank Him for giving people creativity and wisdom to invent. Thank Him for the inventions that give us warmth and light even when we can't be in the sunshine.

UNIT 1 • CHAPTER 4

ACTIVITY 10
# Seeing Created Beauty

Let's enjoy God's beautiful creation outside today! Gather these things to take with you:

1. Spade or shovel to dig with

2. Drawing paper

3. Pencil or crayons

4. Clipboard

5. Camera or phone that can take pictures

6. Garden clippers

7. Flat box or an empty egg carton for collections

Go outside and look for earthworms. You can look under rotting plants or dig a hole in the ground. This is where you will find them eating rotting plants and turning them into good soil for new plants to use. This is the important job God gave them to do! When you find a worm, answer these questions:

ACTIVITY 10: SEEING CREATED BEAUTY

1. Did you find the worm living in a dark or a light place?

   _____

2. Was the worm in a dry or a damp place?

   _____

3. Do you see any eyes on the worm? _____

Go outside to a park, forest, or other area and use your eyes to observe God's creation.

1. Draw or take a picture of a beautiful scene.

2. Make a nature collection. Take time to wander, observe, and collect.

    A. Start a collection of leaves, flowers (only one of each kind), seeds, and seed pods (only one of each kind). These can be dried and pressed by placing them between layers of newspaper and cardboard and putting them under a heavy book. Leave them to dry for two or three weeks. They can also be dried by hanging them upside down in a dry room. Save these to use in Unit 4.

B. Start a collection of dead insects and empty cocoons. Dead butterflies and moths make beautiful specimens. Freezing them for a week should kill any pests that sometimes live inside them. Save this collection for Unit 5.

C. Take pictures or draw pictures of animals, birds, droppings, and other things you can't collect.

# ACTIVITY 10: SEEING CREATED BEAUTY

- Sing the hymn for this unit.

- Repeat the memory verse for this unit.

- Praise God for the beauty you saw today in His creation. Thank Him for giving you amazing eyes that can see it. Thank Him for earthworms and the important job they do.

UNIT 1 • CHAPTER 4

ACTIVITY 11

# Take Care of Your Eyes!

Here is a list of ways our eyes are protected. God protects our eyes in many ways. But we must also protect them. Write "God" after the eye protection God gave us. Write "Me" after the things you must do to protect your eyes.

1. Eyelids that blink to keep things from going into our eyes ⎯⎯⎯⎯

2. Do not look at bright lights ⎯⎯⎯⎯

3. Wear sunglasses when it is too bright outside ⎯⎯⎯⎯

4. Skull bones that protect eyeballs from getting bumped ⎯⎯⎯⎯

5. Do not play with sharp things ⎯⎯⎯⎯

6. Tears to keep eyes from drying out ⎯⎯⎯⎯

7. Pupils that close if light is too bright ⎯⎯⎯⎯

8. When looking at things close to you for a long time, rest your eyes by looking far away every few minutes ⎯⎯⎯⎯

ACTIVITY 11: TAKE CARE OF YOUR EYES!

9. Eyelashes to keep dust out of eyes  _____

10. Eyebrows to keep water and sweat from running into eyes  _____

**Turn away my eyes from looking at worthless things, and revive me in Your way. (Psalm 119:37)**

Sing the hymn for this unit.

Repeat the memory verse for this unit.

Thank God for creating your body to protect your eyes. Pray that He will turn your eyes away from worthless things so you will live in His protection.

UNIT 1 • CHAPTER 4

ACTIVITY 12

# Useful Animal Eyes

O LORD, You preserve man and beast. (Psalm 36:6)

Sit on the floor with both eyes open and ask someone to roll a small ball to you. Try to catch it. Now cover one eye and try to catch the ball. Do you see why it's important for animals that hunt to have two eyes in the front of their head? _____

Take a mirror into a dark room. Let your eyes get used to the dark for a few minutes. Then, while looking into the mirror, switch on the light and watch your pupils. Answer these questions: When you went in the dark room, were you able to see more light after waiting awhile? _____

1. Why?

ACTIVITY 12: USEFUL ANIMAL EYES

2. What happened to your pupils when you switched on the light?

   _____

3. Why? _____

Do you see why bigger pupils help animals see better at night?

_____

Match each animal with its pupil shape.

Rectangle          Tall Slit           Round

Dog pupil          Horse pupil         Cat pupil

53

GOD MADE EVERYTHING

Sing the hymn for this unit.

Repeat the memory verse for this unit.

Thank God for creating people and animal eyes exactly right for each of us.

UNIT 2 • CHAPTER 5
# ACTIVITY 13
# The Big Earth

Go outside and look for the horizon. If you can't see it, watch for it the next time you go somewhere with a lot of open space.

On a globe or world map, find:

1. the North Pole

2. the South Pole

3. the Equator

4. the place where you live

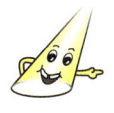

With a marker, draw the equator on your Earth model from Activity 7.

Sing the hymn for this unit.

Write or repeat the memory verse for this unit.

## Memory Verse

"I have come as a light into the world,
That whoever believes in Me
Should not abide in darkness."
(John 12:46)

## ACTIVITY 13: THE BIG EARTH

Thank God for making Earth big enough to fit all the people He made.

UNIT 2 • CHAPTER 5

# ACTIVITY 14
# The Exploding Earth

> [God] touches the hills, and they smoke. (Psalm 104:32)

Let's make a volcano!

## Volcano!

### Ingredients

- A small bottle or narrow jar that can hold 1/4 cup (2 oz. or 60 ml)
- 1/2 teaspoon baking soda (2-1/2 grams)
- Red and yellow food coloring
- 1/4 cup (2 oz. or 60 ml) vinegar
- Water and dirt
- A place to make mud

### Instructions

1. Put the baking soda into the bottle.
2. Put two drops of yellow food coloring and one drop of red food coloring on top of the baking soda. Set the bottle aside.
3. Mix dirt and water to make stiff mud.

ACTIVITY 14: THE EXPLODING EARTH

4. Put the bottle on a flat place and build a mud mountain around it, up to the top of the bottle.
5. You can add toy trees around the volcano if you'd like.
6. Make sure everyone is watching. Then pour the vinegar into the bottle.
7. You can use more of each ingredient if your bottle is bigger.

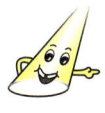

Sometimes farmers live near volcanoes because the dirt there is very good for growing food. Would you like to live near a volcano?

_____

Sing the hymn for this unit.

Repeat the memory verse for this unit.

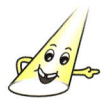

Ask God to protect the people that live near volcanoes. Pray that they will get warned early enough so they can leave before it erupts.

UNIT 2 • CHAPTER 5
ACTIVITY 15
# The Pulling Earth

What holds everyone on the earth? Gravity or glue?

_____

Color the picture.

Sing the hymn for this unit.

Repeat the memory verse for this unit.

Thank God for making our Earth just the right size for life. Thank Him that we don't float off and that air stays here for us to breathe.

ACTIVITY 15: THE PULLING EARTH

# GOD MADE EVERYTHING

UNIT 2 • CHAPTER 6
# ACTIVITY 16
# Rock Hunt

Read Matthew 7:24-27. Draw lines to match the builder to his house. Circle the one who obeys Jesus.

Wise

Foolish

GOD MADE EVERYTHING

Get an empty egg carton that held a dozen eggs. How many is a dozen? _____. Go outside and find a dozen rocks to put in your egg carton. Try to find rocks that are different from each other.

1. How many rocks look like they are made of sand or mud? _____

2. How many rocks have sparkles? _____

3. How many different colors of rocks did you find? _____

Now let's plant some seeds! Some will be in rock, and some will be in dirt like in Jesus' story from Luke 4:4-15. You will need some dried beans for this activity.

### Seed Experiment

1. In your egg carton, move all the rocks you collected to a few sections on one side. The number of sections for rocks will depend on how big your rocks are. If your carton is made of plastic or Styrofoam you will need to poke a 1/4 inch (0.5 cm) drainage hole in the bottom of each section.
2. Put dirt in the sections without rocks.
3. Plant a bean in each section. Bury it about one inch (2.5 cm) deep in the dirt or rocks.

ACTIVITY 16: ROCK HUNT

4. Water the dirt enough to make it damp. Water the rocks the same amount.
5. Place your carton on a tray in a bright window.
6. Water the dirt a little each day to keep it damp. Water the rocks the same amount.
7. Check on your carton each day to see what happens. Keep checking on it until Unit 4.

Sing the hymn for this unit.

Repeat the memory verse for this unit.

Thank God for Earth's strong crust to build our houses on.

UNIT 1 • CHAPTER 6

ACTIVITY 17

# Mine for Metal

Miners have to know how to separate metal from rock. Let's see if we can separate iron from dirt! Iron can be pulled to magnets the same way we are pulled to Earth by gravity. You will need a small magnet and a small plastic bag.

Copper mine

## Iron from Dirt

1. Put the magnet in the plastic bag and take it outside to some dry dirt.
2. Push the bagged magnet around in the dirt. If black fuzz is collecting around the magnet, you have found iron!
3. Turn the bag inside out to separate the iron from the magnet.

Sing the hymn for this unit.

ACTIVITY 17: MINE FOR METAL

Repeat the memory verse for this unit.

Thank God for all the useful things made of metal!

Surely there is a mine for silver,
And a place where gold is refined.
Iron is taken from the earth,
And copper is smelted from ore. . . .
As for the earth, from it comes bread,
But underneath it is turned up as by fire;
Its stones are the source of sapphires,
And it contains gold dust.
(Job 28:1, 2, 5, 6)

A sapphire jewel

UNIT 1 • CHAPTER 6

ACTIVITY 18

# Make a Flood and Fossil Frittata

Ask someone to read to you the account of the worldwide flood in Genesis 6:6-8:4.

Now let's make an edible model to learn about the flood.

## Flood and Fossil Frittata Ingredients

**Ark:**

- 1 hot dog bun
- 1/4 cup (60 g) butter (softened)

**Animals and People:**

- A small handful of tiny seeds (cumin, poppy, quinoa, or chia)

**Fountains of the deep:**

- 16 oz (450 g) grated cheddar cheese
- 1 10.5 oz (298 g) can cream of chicken soup

**Earth's crust:**

- 3 oz (80g) tortilla chips

**Dirt:**

- 1/4 cup (40 g) corn meal

## ACTIVITY 18: MAKE A FLOOD AND FOSSIL FRITTATA

**Plants:**

- 8 oz (230 g) can chopped green chiles

**Rain:**

- 6 eggs, beaten
- 1 cup (0.25 L) milk
- 1/2 teaspoon (3 g) salt

**New Life:**

- Lettuce, finely shredded
- Diced, fresh tomatoes

### Flood and Fossil Frittata Instructions

**For the ark:**

1. Spread a thin layer of butter on the inside of the bun.
2. Sprinkle some seeds on the butter to be Noah's family and the animals God brought to the ark.
3. Close the bun and spread the rest of the butter all over the outside of the bun to show that the ark was covered with pitch. Put the ark in the freezer to harden the pitch.

GOD MADE EVERYTHING

**For the Land:**

1. Sprinkle the cheese on the bottom of a 9"x13" (23 cm x 33 cm) baking pan.
2. Spread the canned soup over the cheese. This is the "fountains of the deep" under the crust.
3. Lay tortilla chips side by side to cover the soup. Press a few down to see the fountains of the deep coming up.
4. Sprinkle cornmeal on the chips for dirt. Put the ark on the dirt.
5. Cover the cornmeal with chiles. The chiles will be plants and trees.
6. Sprinkle the rest of the seeds on the land to be animals.

**For the Flood:**

1. Mix the eggs, milk, and salt.
2. Pour the mixture over the land and the ark, making it rain. Count to forty, the number of days it rained.
3. Tip, swirl, and stir the pan until everything is well mixed. Keep the ark safe like God did with the real ark.

**The Ark Comes to Rest:**

1. Bake at 400 degrees (200 C) for 30 minutes, or until eggs are set.
2. Before you eat, slide a spatula under the frittata and lift to show the mountains of Ararat forming. Notice the buried fossils of plants and animals.

ACTIVITY 18: MAKE A FLOOD AND FOSSIL FRITTATA

3. Sprinkle lettuce and tomatoes on top to show that God made new life on the earth.

**Eat!**

Sing the hymn for this unit.

Repeat the memory verse for this unit.

Thank God for the beautiful mountains that were formed when the crust moved at the time of the worldwide flood. When you see fossils, thank God for keeping Noah, his family, and the animals safe to live on the earth after the flood.

UNIT 2 • CHAPTER 7

# ACTIVITY 19
# Peace; Be Still!

- Let's make waves! Fill a cup, pan, or bucket with water. Blow on one side to see the waves your breath can make.

- Draw a cartoon character of wind (like Beamer's friend Puff, if you like) blowing on the wave on the coloring page. Color the picture.

We read in Psalms that God rules the sea and makes the waves be still. Many years after the psalms were written, Jesus did that! He was in a boat with His friends when a wind storm came and the waves were filling up the boat. Jesus was asleep on a pillow. His friends were afraid, so they woke Him. Jesus said to the sea, "Peace; be still!" and the wind and water became calm. He reminded His friends to not be afraid and to have faith in Him.

- Sing the hymn for this unit.

- Repeat the memory verse for this unit.

- Ask God to help you not be afraid. Pray that He will give you faith to trust Him even in times of trouble.

ACTIVITY 19: PEACE; BE STILL!

# GOD MADE EVERYTHING

UNIT 2 • CHAPTER 7
ACTIVITY 20
# Water for Life

More than half of your body is made of water! God makes you feel thirsty so that you will drink enough to keep your body healthy. Water has a lot of jobs to do inside you.

The water that comes out when you use the toilet is called urine. It is yellow because of waste that is being cleaned out of you. If it is dark colored or if you do not have to use the toilet very often, you should drink more water.

There are other ways water can leave your body. Let's see if we can find them! You will need a small mirror, a spoon, and a tissue.

Look at your face in the mirror with your mouth closed. What two places can you see wetness?

_____

That wetness is tears. Tears keep your eyes from drying out.

Put the mirror in your refrigerator for 15 minutes. While you are waiting, lick the spoon. That wetness is saliva. God gave you saliva to help you chew and swallow dry food.

Hold the tissue loosely over your nose. Sniff air in and out a few times. The wetness you feel in your nose is mucus. It helps catch dirt from the air you breathe, and keeps your nose from drying out inside.

Take the mirror out of the refrigerator. Cup your hand over it for a few seconds. When you lift your hand, look quickly at the mirror. Do you see vapor fogging the mirror where your hand was? _____ If not, try using your foot. If that doesn't work, run around the house and try the experiment again. God makes water come out of your skin to keep you cool. This water is called sweat.

ACTIVITY 20: WATER FOR LIFE

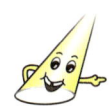

Before the mirror warms up, breathe on it with your nose first and then with your mouth. This vapor is coming from inside you when you breathe. It is not saliva and not mucus.

Sing the hymn for this unit.

Repeat the memory verse for this unit.

Praise God for wonderful water! Give thanks for this special liquid that does so many things in our bodies.

Jesus once told a woman that even though people drink, they get thirsty again. But Jesus can give eternal life that is like a fountain of water that never ends!

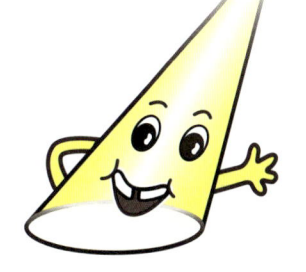

UNIT 2 • CHAPTER 7

ACTIVITY 21

# Cycle Some Water

**You visit the earth and water it. (Psalm 65:9)**

Let's see how the sun's heat makes vapor go up! You will need a small, self-sealing plastic bag, blue food coloring, an empty glass, and a sunny window.

## Water Vapor Experiment

1. Put about an inch (2.5 cm) of water in the plastic bag without getting the sides wet.
2. Add a drop of blue food coloring to the water.
3. Close the bag so it is tightly sealed.
4. Lean the bag against a sunny window. Use the empty glass to hold the bag in place.
5. Check the bag after an hour. Do you see little drops of water on the sides of the bag? This is water vapor that has come up from the water in the bottom of the bag. The water on the side of the bag should be clear. The blue coloring has been left behind just as salt is left behind in the ocean.
6. Tap the sides of the bag to make it rain.

ACTIVITY 21: CYCLE SOME WATER

- Sing the hymn for this unit.

- Repeat the memory verse for this unit.

- Praise God for His way of making water do its jobs and cycle around to do them again and again.

UNIT 2 • CHAPTER 8

**ACTIVITY 22**

# Happy Birthday! (to Someone)

How old are you now?

_____

How old were you last year at this time?

_____

How many days will there be until next year at this time?

_____

Draw arrows on the dotted line of the traveling earth picture to show the Earth going around the sun in a counterclockwise direction.

Let's act out a year! Find a helper or a ball to be the sun. Put the sun in the middle of an open space. Pretend you are the earth. Walk around the sun counterclockwise. Do it again, this time spinning counterclockwise at the same time. Are you dizzy?

ACTIVITY 22: HAPPY BIRTHDAY! (TO SOMEONE)

Sing the hymn for this unit.

Repeat the memory verse for this unit.

Thank God that birthdays happen once a year!

UNIT 2 • CHAPTER 8

ACTIVITY 23

# Seasons of the Year

For this activity, you will need your Styrofoam model of the earth from Activity 7 or a round fruit and skewer.

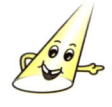

Practice holding your earth tilted this way as shown in this picture.

23.5°

Take your model to a dark room. Take along someone with a flashlight. Hold your earth at the tilted angle you were practicing and walk around the flashlight sun. Don't try to spin your earth for this activity. Be sure to keep the earth's tilt the same while you walk. Your helper will need to keep turning in order to keep the light shining on the earth. Notice that part of your way around the sun, the bottom half of the earth has the light shining directly on it. Keep walking and notice how the sun shines directly on the top half of the earth when you are on the other side of the circle.

Near the equator, the seasons do not change like they do farther away from it. That's because the sun shines on the equator the same way all year long.

In places where seasons do change, the weather changes too. Some trees also change. They flower in spring, make fruit and

ACTIVITY 23: SEASONS OF THE YEAR

seeds in summer, lose leaves in fall, and rest in winter.

Children enjoy different things in each season. On the seasons page, under each picture, write the season it shows. Under that, draw a food you like to eat in each season.

While the earth remains,
Seedtime and harvest,
Cold and heat,
Winter and summer,
And day and night
Shall not cease.
(Genesis 8:22)

Sing the hymn for this unit.

Repeat the memory verse for this unit.

Praise God for the beauty of the changing seasons. Thank Him for something you like to do in each season.

GOD MADE EVERYTHING

# Summer, Fall, Winter, Spring

84

UNIT 2 • CHAPTER 8

# ACTIVITY 24
# Hibernate!

It must be cozy to hibernate. Let's pretend you are a furry animal and winter is starting.

- Make a den out of blankets. You can hang the blankets over chairs, clothes baskets, or boxes. Put a pillow inside.

- Eat a snack so you won't get hungry while hibernating.

- Go into your den and lie down. Try not to move. Maybe you will fall asleep.

- Sing the hymn for this unit.

- Repeat the memory verse for this unit.

- Thank God for caring for animals and for you when it's cold.

UNIT 3 • CHAPTER 9

## ACTIVITY 25
# Your Five Senses

God gave your body five ways to tell what is going on around you: seeing, hearing, feeling, tasting, and smelling. These are your **senses**.

Are you are able to sense light and air (wind) with each of your senses? Write "Yes" or "No" in the Beamer and Puff chart below.

| Can you... | Beamer (Light) | Puff (Air) |
|---|---|---|
| See? | | |
| Hear? | | |
| Feel? | | |
| Taste? | | |
| Smell? | | |

Sing the hymn for this unit.

ACTIVITY 25: YOUR FIVE SENSES

Write or repeat the memory verse for this unit.

> **Memory Verse**
>
> The darkness and the light are both alike to You.
> (Psalm 139:12)

Thank God for giving us air to breathe. You don't have to buy it, cook it, or carry it around (unless you are scuba diving). Thank Him that you can feel yourself pushing air when you run.

UNIT 3 • CHAPTER 9
ACTIVITY 26
# How to Put Out a Fire

Fire needs three things to get started and to keep burning. A fire will go out if one of these is taken away:

1. Fuel: Without fuel, a fire can't burn. When a fire starts in a forest, firefighters try to make a bare, open line of ground with nothing growing on it. When the fire reaches this line, it can't burn any further because it has run out of fuel.

2. Heat: Firefighters are glad when cool weather comes because it slows down fires and makes it harder for new ones to start.

3. Oxygen: Firefighters use water and foamy spray to block oxygen from the burning fuel and to cool it. Small fires can be put out by covering them with dirt. If you have a fire in a cooking pan, you can put the fire out by putting a lid on the pan to block the oxygen from getting to the fire.

Let's do an activity to help us remember that fire needs air to burn.

### ACTIVITY 26: HOW TO PUT OUT A FIRE

## Put Out a Fire

1. Put a tea light or votive candle in a metal pan.
2. With an adult's help, light the candle.
3. Put a glass jar over the burning candle. What happened to the flame? When fire runs out of oxygen, the fire has to stop burning.
4. Remember that fire is dangerous. Never play with fire!

There is one kind of fire that we always need to keep burning! Jesus talks about how we need to shine by doing good to others. Those that see this will glorify God:

**Let your light so shine before men, that they may see your good works and glorify your Father in Heaven. (Matthew 5:16)**

Sing the hymn for this unit.

Repeat the memory verse for this unit.

Thank God for heat from fire. Thank Him for ways to put out fire.

UNIT 3 • CHAPTER 9

# ACTIVITY 27
# Wing It!

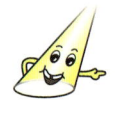

Let's learn about wings!

## Make a Wing Shape

1. Gather a stapler and a cardboard tube left from a paper towel or toilet paper roll.
2. Pinch the tube together to make a flat, back edge. Leave the front rounded.
3. Staple along the flat edge to keep it flat.
4. This is the shape of a wing. Wings make it possible for birds or airplanes to rise up off the ground and fly. Is your cardboard wing strong enough to let you fly?

Sing the hymn for this unit.

Repeat the memory verse for this unit.

Praise God for the gift of flight. Thank Him for flying birds and for airplanes that allow people to fly too!

UNIT 3 • CHAPTER 10

ACTIVITY 28
# Five Layers

Our sky has five different layers in it. To learn more about layers, let's make some pumpkin pie spice!

## Pumpkin Pie Spice

**Ask an adult to let you use an empty clear glass spice jar or some other small, narrow jar. Gather the following spices:**

- 3 tbsp. ground cinnamon (24g) (to be the earth)
- 2 tsp. sugar (4g) (first layer of sky)
- 2 tsp. ground ginger (4g) (second layer of sky)
- 2 tsp. nutmeg (5g) (third layer of sky)
- 1 ½ tsp. ground cloves (3g) (fourth layer of sky)
- 1 ½ tsp. ground allspice (3g) (fifth layer of sky)

**Instructions:**

1. Using a funnel, pour each spice into your jar, one at a time. Wipe off any dust on the sides of the jar by running a dry finger inside the jar before adding another layer. Gently tap the jar on a table to flatten each layer before adding the next.
2. When you finish adding all the spices to the jar, count your layers. Do you have one layer for the earth and five for the sky?

GOD MADE EVERYTHING

> 3. Now pour the layers into a bowl and mix them up with a spoon. Use the funnel to pour them back into the jar for storage. Use in pumpkin pie, pumpkin muffins, warm milk, or any recipe calling for pumpkin pie spice.

Find a swing you can use. Practice pumping your legs and leaning back and forth on your arms to make yourself swing high. Ask someone to read you this part of a poem by Robert Louis Stevenson while you are swinging:

ACTIVITY 28: FIVE LAYERS

**The Swing**
How do you like to go up in a swing,
Up in the air so blue?
Oh, I do think it the pleasantest thing
Ever a child can do!

Sing the hymn for this unit.

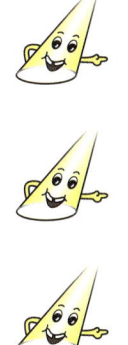

Repeat the memory verse for this unit.

Thank God for protecting us with our beautiful blue sky's layers.

UNIT 3 • CHAPTER 10
# ACTIVITY 29
# Cotton Clouds

There are three main kinds of clouds:

Cirrus, feathery like floating pillow feathers

Stratus, flat and big like a soft blanket

Cumulus, like puffy cotton balls

94

ACTIVITY 29: COTTON CLOUDS

Let's make these three kinds of clouds!

1. Gather a sheet of blue paper, some cotton balls, and glue.

2. Tear, pull, and shape the cotton to look like each cloud type.

3. Glue your clouds on the paper and draw anything else you like in your sky.

Sing the hymn for this unit.

Repeat the memory verse for this unit.

God made such beautiful clouds! Praise Him for all the different shapes of clouds and how they make sunrises and sunsets so beautiful!

UNIT 3 • CHAPTER 10
ACTIVITY 30
# Snowflake Art Project

It takes an hour for a snowflake to fall to the ground from a cloud. Let's make snowflakes!

- Gather a pair of scissors and one sheet of white paper for each snowflake you would like to make.

- Follow the instructions on the next page.

- Sing the hymn for this unit.

- Repeat the memory verse for this unit.

- Thank God for beautiful snowflakes. Snowflakes are another way He gives us water!

# ACTIVITY 30: SNOWFLAKE ART PROJECT

## HOW TO MAKE PAPER SNOWFLAKES

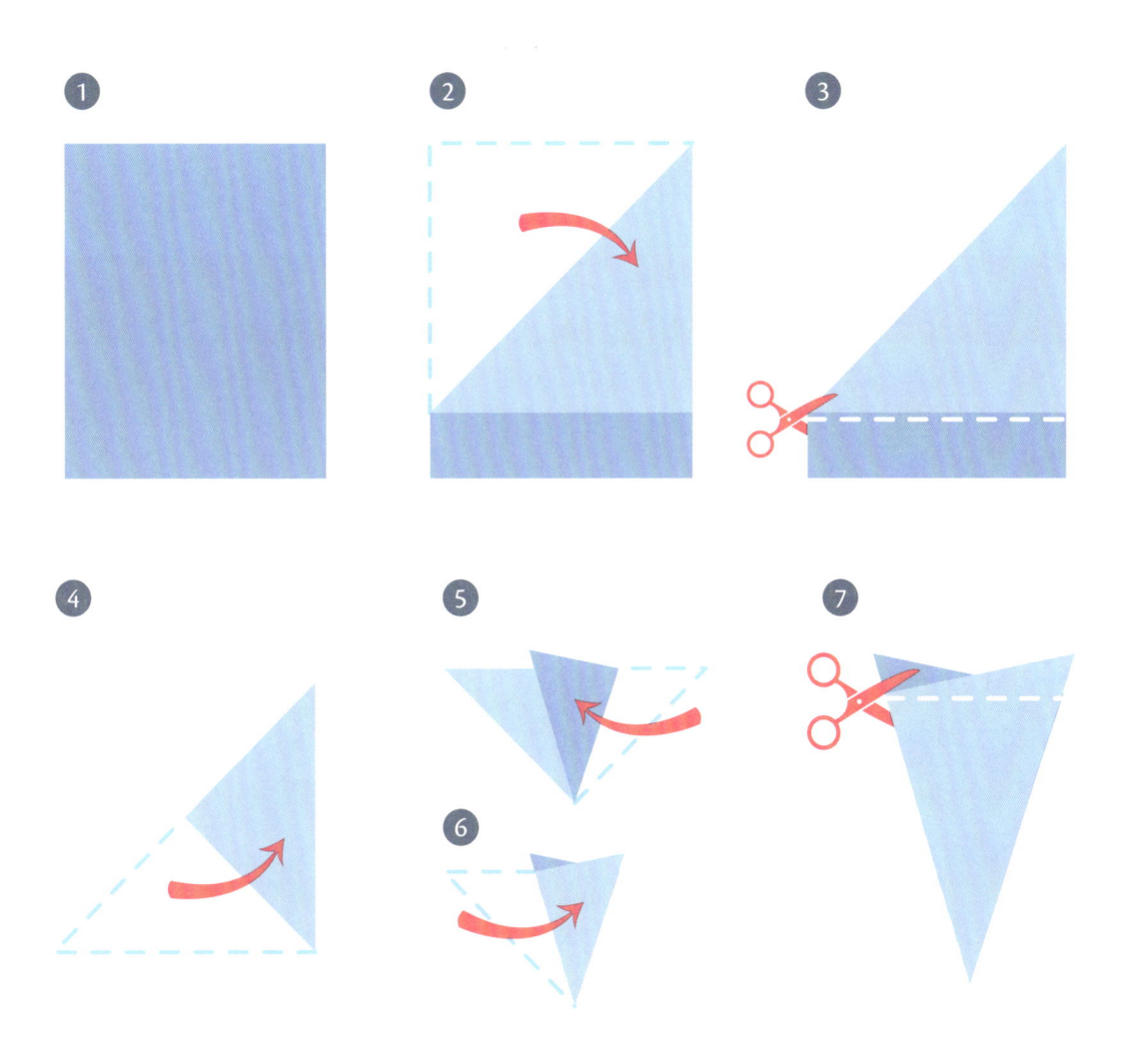

USE YOUR IMAGINATION! SHAPE YOUR SNOWFLAKE BY CUTTING IT IN DIFFERENT WAYS!

UNIT 3 • CHAPTER 11

ACTIVITY 31

# Man on the Moon

### Definition

An **astronaut** is a person who travels past the earth's sky into outer space.

Did you know that people have gone to the moon in rocket ships? They had special suits and food for the trip. They had to take air and water along. They took science tools to help them learn about the moon. They brought back moon rocks! Most moon rocks are kept at the Lyndon B. Johnson Space Center in Houston, Texas.

- Color the astronaut picture. Ask an adult to find "Claire de Lune" (Moonlight) by Claude Debussy for you to listen to as you color.

- Sing the hymn for this unit.

- Repeat the memory verse for this unit.

ACTIVITY 31: MAN ON THE MOON

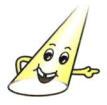

Thank God that He cares for us and gives us the moon for a nightlight. Praise Him for the moon's job of keeping the earth tilted.

99

# GOD MADE EVERYTHING

UNIT 3 • CHAPTER 11
## ACTIVITY 32
# Eclipse of the Sun

Let's learn what the moon does to make an eclipse of the sun:

1. Gather a flashlight, some tissues, and a marble you can't see through. Take them to a dark room.

2. Wrap enough tissues around the flashlight so the light is not too bright to look at. Set it up at the level of your head. It will be the sun.

3. The marble will be the moon, and your eyes will be the earth. Stand about an arm's length away from the sun and hold the marble between your eyes and the lit flashlight.

4. Move the "moon" forward and backward between the earth and sun until it looks like the moon exactly covers the sun. This is a total eclipse of the sun.

5. Now move the moon slowly sideways to see a different kind of eclipse that does not cover the sun all the way. When the marble only partially covers the light from the flashlight sun, it seems to take a bite out of it. Can you make your eclipse look like the pictures on the next page?

GOD MADE EVERYTHING

6. If you can, leave the flashlight set up for the next activity.

- Sing the hymn for this unit.

- Repeat the memory verse for this unit.

- Thank God for putting the moon where it will exactly cover the sun during an eclipse. It shows that He is thinking of us!

UNIT 3 • CHAPTER 11

# ACTIVITY 33
# Moon Shapes

> He appointed the moon for seasons; the sun knows its going down. (Psalm 104:19)

The lit part of the moon is what we see at night. It seems to change shape during the month. Let's see why this is:

1. Take a flashlight and a ball that is not shiny (like a tennis ball or baseball) to a dark room.

2. Set the flashlight up at the level of your head.

3. The flashlight is the sun, the ball is the moon, and your head is the earth. Stand in front of the flashlight but far enough away that you don't touch it. Hold the moon out at arms length in front of you. Be sure to stand close enough that you can see light on the moon when you turn in a circle.

4. Start with the sun behind you and the moon in front. Hold the moon high enough so your head is not shading it. Your moon is all lit up. This is a full moon.

103

GOD MADE EVERYTHING

5. Slowly spin toward the sun, watching for the lit part of the moon to change to a half moon and then a crescent moon.

6. Turn more until the moon is blocking the sun. (That is a total eclipse of the sun.) Lower the moon a little to see that it is all dark. This is a new moon.

Now we see that the moon looks like it changes shape because it is moving around the earth. We are seeing different amounts of the moon lit up when it is in different places.

Sing the hymn for this unit.

Repeat the memory verse for this unit.

Praise God for the interesting moon and all its shapes!

UNIT 3 • CHAPTER 12
ACTIVITY 34
# The Lonely Little Space Rock

*Let me tell you a story I thought of to help you learn about space rocks!*

### The Lonely Little Space Rock
*By Beamer*

As a bouncy beam of light, I usually do my job on Earth. But one day I decided to take a trip into the solar system. I found an Earth mountain to bounce off of and I flew into space. Suddenly, I found a bright, beautiful, white **comet**. Its head was a big ball of ice. It had a vapor tail that stretched far into space.

Then I saw a little space rock frozen onto the comet's head. The rock was trying with all its strength to get unstuck.

"What are you doing, little space rock?" I asked.

"I am trying to get out!" he answered. "You see, I love being in the heavens that show God's glory, but I have been in this comet for thousands of years and I'm very lonely. I would like to be free and find some other rocks to play with."

# GOD MADE EVERYTHING

"Well," I said, "see that vapor tail coming off your comet? That means that you have come close enough to the sun for some of the ice to melt, and—"

Just then, some of the ice around the little space rock turned into vapor and joined the comet's tail! The little space rock pushed himself loose. He was now a **meteoroid**!

"Yay! God has made me free!" he shouted, and the little space rock waved goodbye and began to travel around the sun.

A long time later, when I had bounced into the solar system again, I found the little space rock. I was happy to see him and he was happy to see me. He was glad to be free, but still seemed a little lonely.

He asked me, "What is that pretty blue ball you just came from?"

"That," I said, "is Earth. It is the most special place in all of God's creation. It has life!"

I tried to explain life, but I finally had to say that only God could really understand it.

"Are there any rocks on Earth?" he asked.

"Oh, yes!" I said, "lots of different kinds that you have never seen before."

"Do you think we could go there?"

"Well, if your path around the sun takes you close enough to the earth, its gravity might pull you in, but—" I wanted to warn him that most space rocks burn up in the third layer of Earth's sky. But suddenly, he began to move away from me faster and faster. He was scared! The earth's gravity was pulling the little space rock toward it!

## ACTIVITY 34: THE LONELY LITTLE SPACE ROCK

"Ow! Ow! Ow! Ow! Ow!" I heard. I saw that the little space rock was on fire! He had become a **meteor**! Quickly, I bounced off the moon and made it to the earth before him. I saw him land on the nighttime side of the earth in some soft, cool mud. Now that he had landed, he was a **meteorite**.

"Ahhh," said the little space rock as he sizzled. "That was hot!"

"Well, you sure looked beautiful coming down!" I said. "Of course, I always like to see things that give off light."

Soon the earth turned, and more light joined us from the sun. The little space rock looked around, amazed at all the life, and at all his soon-to-be rock friends.

"Wow! What a beautiful place!" he said with tears coming to his eyes. "I have been lonely, and I have been scared and hurt. But when I see this beautiful earth and know that God brought me to it, I can only say, 'God is good!'"

Choose from the following words to answer the questions below:

Meteor    Meteorite    Meteoroid    Comet

1. When the little space rock was stuck for thousands of years, what was he frozen onto?

_____

107

# GOD MADE EVERYTHING

2. When the little space rock was traveling around the sun like the comet he had come off of, he was a

_____ .

3. When the poor little space rock was on fire in Earth's third layer of sky, he was a

_____ .

4. When the little space rock was in the cool mud of Earth, he was a

_____ .

Sing the hymn for this unit.

Repeat the memory verse for this unit.

Praise God for His beautiful earth, glorious heavens, and His goodness to you.

108

UNIT 3 • CHAPTER 12
ACTIVITY 35
# Jump to the Planets!

MOON
MERCURY
VENUS
EARTH
MARS
JUPITER
SATURN
URANUS
NEPTUNE
SUN

Find a place where you have room to jump forward 10 times in a row. This could be a sidewalk you can mark with chalk, dirt you can mark with scratched lines, or a floor you can mark with removable tape.

Starting at one end, mark a place to start (the sun). Make eight more marks, one jump's length apart, for the planets.

Stand at the sun. Roll a die and take that many jumps forward. Look at the order of the planets in the list below and see which planet you landed on. Act out that planet using the ideas here or some of your own. Keep rolling the die and

109

jumping forward and backward until you have been to all the planets.

1. Mercury: Act hot and then cold while spinning.

2. Venus: Pretend to breathe poisonous air and act very hot.

3. Earth: Act happy and comfortable.

4. Mars: Cough and brush off dust.

5. Jupiter: Run away from the always-moving storm.

6. Saturn: Sweep your hand across the sky to show rings.

7. Uranus: Do a summersault to show its funny spin.

8. Neptune: Shiver with cold and try to see the far-away sun.

Sing the hymn for this unit.

Repeat the memory verse for this unit.

Praise God for the interesting planets that are each different from each other.

UNIT 3 • CHAPTER 12

# ACTIVITY 36
# Draw a Star

Do the dot-to-dot! Start at the number 5 and draw a line to the number 1. Keep drawing lines to the numbers in order until you have made a star!

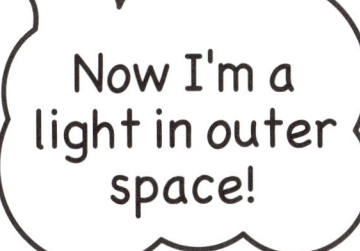

• 1

• 3                                        • 4

• 5                                        • 2

Now I'm a light in outer space!

111

GOD MADE EVERYTHING

God sent a star to show wise men the way to find Jesus:

**And behold, the star which they had seen in the East went before them, till it came and stood over where the young Child was. When they saw the star, they rejoiced with exceedingly great joy. (Matthew 2:9-10)**

Sing the hymn for this unit.

Repeat the memory verse for this unit.

Praise God for using a star to lead the wise men to His Son Jesus!

UNIT 4 • CHAPTER 13

ACTIVITY 37
# Roots Store Food

Roots are a plant's cupboard or pantry. They are a place to save food for later. Sometimes a plant gets eaten above ground. Sometimes the parts above ground die in the winter. When this happens, a plant's leaves can often start growing back by using food stored in its roots.

Let's see if you can get a plant to grow again from its roots!

## Grow It Again

1. Gather a shallow dish, an adult with a sharp knife, and a few root vegetables like carrot, beet, or turnip.
2. Ask the adult to cut one inch (2.5 cm) off the top of your roots.
3. Set the root tops in the pan and add enough water to make it about 1/2 inch (1 cm) deep.
4. Put the pan near a bright window.
5. Check on your root tops every day and add water when needed.
6. It may take a few weeks for the leaves to grow.
7. When leaves start growing, you can put a little boat in the water and land it on your pretend islands.

GOD MADE EVERYTHING

Sing the hymn for this unit.

Write or repeat the memory verse for this unit.

### Memory Verse

Let the field be joyful, and all that is in it.

Then all the trees of the woods will rejoice before the LORD.

For He is coming.

(Psalm 96:12-13)

ACTIVITY 37: ROOTS STORE FOOD

Thank God for the food stored in roots that can let plants grow again. Thank Him for a root food that you like to eat.

UNIT 4 • CHAPTER 13

## ACTIVITY 38
# Watch a Plant Drink

Let's experiment to see what the thin tubes in plants can do!

## Thirsty Plant

1. Gather a clear glass or jar, a stalk of celery, some blue or red food coloring, and an adult with a sharp knife.
2. Fill the glass about a third full of water.
3. Add three or four drops of food coloring.
4. Ask the adult to make a fresh cut at the bottom of the celery.
5. Put the celery in the liquid right away.
6. Look at the celery every fifteen minutes over the next few hours. Look for colored lines in the stem to see the special tubes. Watch for the leaves to slowly change color.

ACTIVITY 38: WATCH A PLANT DRINK

- Observe your egg carton bean plants you planted earlier. Look up Luke 8:6. Do any of your bean plants look like the plants in this verse?

- Sing the hymn for this unit.

- Repeat the memory verse for this unit.

- Thank God for the tiny tubes that bring water and minerals upward in a plant and take food downward.

UNIT 4 • CHAPTER 13

ACTIVITY 39
# Gardening

Have you ever grown or taken care of plants outside? If so, you were gardening! Farmers grow a lot of food to sell to others. It's their job. But you can grow a small amount of food for your own family. Your garden will need a patch of ground that gets at least six hours of sun a day. You will also need some seeds and water to get started. Just follow the instructions on your seed packages. Gardening is a great way to enjoy God's creation while you grow some healthy food!

---

**For as the earth brings forth its bud,
As the garden causes the things that are sown in it to spring forth,
So the Lord GOD will cause righteousness and praise to spring forth before all nations.
(Isaiah 61:11)**

---

Have fun helping the little pea find its way back into the pod!

Sing the hymn for this unit.

Repeat the memory verse for this unit.

ACTIVITY 39: GARDENING

Praise God that people can grow food instead of having to look for it like animals do.

119

UNIT 4 • CHAPTER 14

ACTIVITY 40
# Plant Some Seeds

It's fun to plant seeds and watch them sprout! God put life in each little seed. He sends His warmth and light to make the seeds grow. All you have to do is get the seeds started and keep them watered. God does the rest!

Gather these things:

1. Pots, egg cartons, or plastic containers to plant seeds in

2. Potting soil or dirt

3. Seeds you find from food (avocados, grapefruit, whole grain, popcorn, dried beans, and dill seeds or cumin seeds from your spice cupboard) or seeds you buy from a gardening store. Radish seeds will come up in about a week. Seeds from trees take a very long time to sprout but have beautiful leaves when they do.

4. With an adult's help, poke holes in the bottom of your containers if needed. Seeds will rot if their dirt stays too wet.

5. Fill each pot with soil until it's about an inch (2.5 cm) from the top.

ACTIVITY 40: PLANT SOME SEEDS

6. Make a hole in the dirt with your finger or a spoon. The hole should be twice as deep as the size of your seed.

7. Plant the seed in the hole and cover it with dirt.

8. Put your pots near a sunny window.

9. Water the dirt so it's moist. Add more dirt if needed. Check on the soil each day and add more water as needed to keep it moist. You can put plastic wrap over the pots to keep the dirt from drying out. Be sure to take it off when your plants sprout.

---

Then God said, "Let the earth bring forth grass, the herb that yields seed, and the fruit tree that yields fruit according to its kind, whose seed is in itself, on the earth"; and it was so. (Genesis 1:11)

---

Sing the hymn for this unit.

Repeat the memory verse for this unit.

Praise God for all the different seeds He put in the world. Ask Him to help you remember to water your seeds.

UNIT 4 • CHAPTER 14

ACTIVITY 41

# Stretch Like a Stem

Pretend to be a sunflower stem. Hold your arms above your head. Stretch one of your arms as high as you can. Notice how the other side of you bends like a stem reaching toward the light!

Now stand by a window or outside in the sun. Stretch your shady side. That should make you lean toward the light!

Vines lean toward things they are touching. This makes them wrap around trees and poles. Pretend to be a vine! Find something you can wrap around by curling where you touch it.

The trunks and branches of trees are their stems. Tree stems are covered with bark to protect them. Different kinds of trees have different kinds of bark. Color the trees on the coloring page with different colors of bark and leaves.

Sing the hymn for this unit.

Repeat the memory verse for this unit.

Thank God for stretching. It helps plants reach the sunlight. Stretching helps you reach things too!

ACTIVITY 41: STRETCH LIKE A STEM

123

# GOD MADE EVERYTHING

UNIT 4 • CHAPTER 14
ACTIVITY 42
# Let's Play Pollination!

A pollinator is like a postal worker who delivers your mail. Pollinators deliver information. A postal worker gets paid money for doing his job. A pollinator gets nectar for doing its job.

Let's pretend to be pollinators. You are going to help bring together pollen and eggs.

1. Gather crayons, scissors, tape, four envelopes, and two glasses of water or juice for nectar.

2. The next two pages are the same. They each show half of an avocado fruit to color. On each picture, color the yummy part of the avocado light green. Color the skin dark green and the seed brown.

3. Pollen is shaped differently in different kinds of plants. Avocado pollen looks like a spiky ball.

4. Cut each picture in half on the dotted line.

5. Put each picture half in its own envelope. Write an "E" on the outside of the envelopes that have an egg picture inside. Write a "P" on the envelopes that have a pollen picture inside.

# GOD MADE EVERYTHING

6. Find two rooms to be flowers. In each room, put an "E" envelope, a "P" envelope, and a glass of nectar.

7. Now fly around the house looking for nectar.

8. When you visit a flower room, take its pollen envelope to the other flower room. Drink some nectar!

9. Pick up that flower's pollen envelope and take it back to the first flower room. Drink some more nectar!

10. Open the envelopes and tape each egg half together with a pollen half.

11. Now two avocado seeds have all the instructions for making two avocado trees!

Sing the hymn for this unit.

Repeat the memory verse for this unit.

Thank God for pollen and what it does!

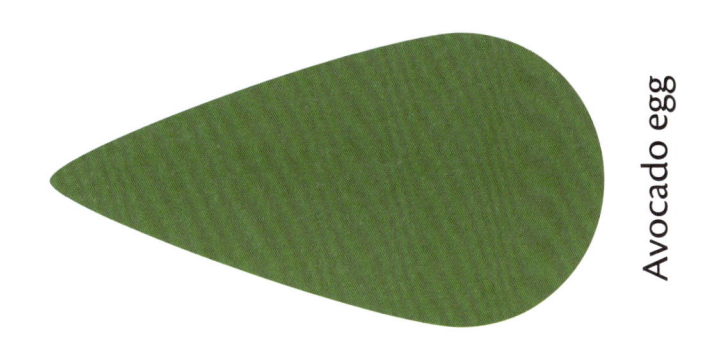

## ACTIVITY 42: LET'S PLAY POLLINATION!

Avocado egg

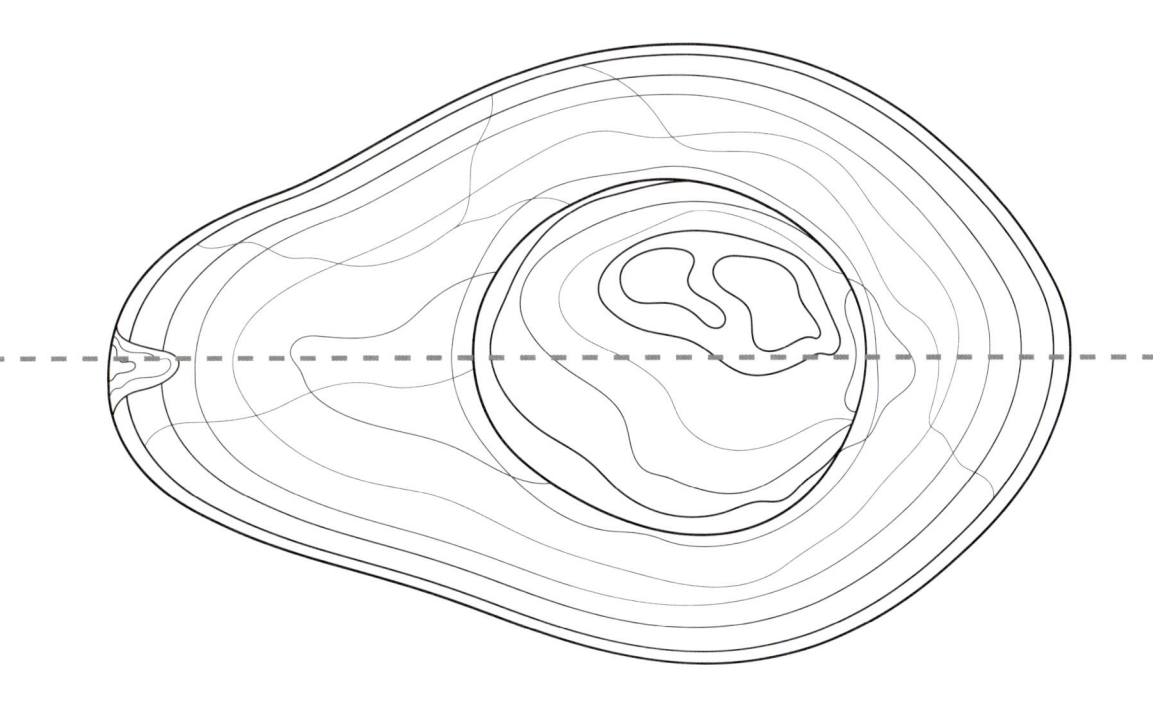

Avocado pollen

127

# GOD MADE EVERYTHING

# ACTIVITY 42: LET'S PLAY POLLINATION!

Avocado egg

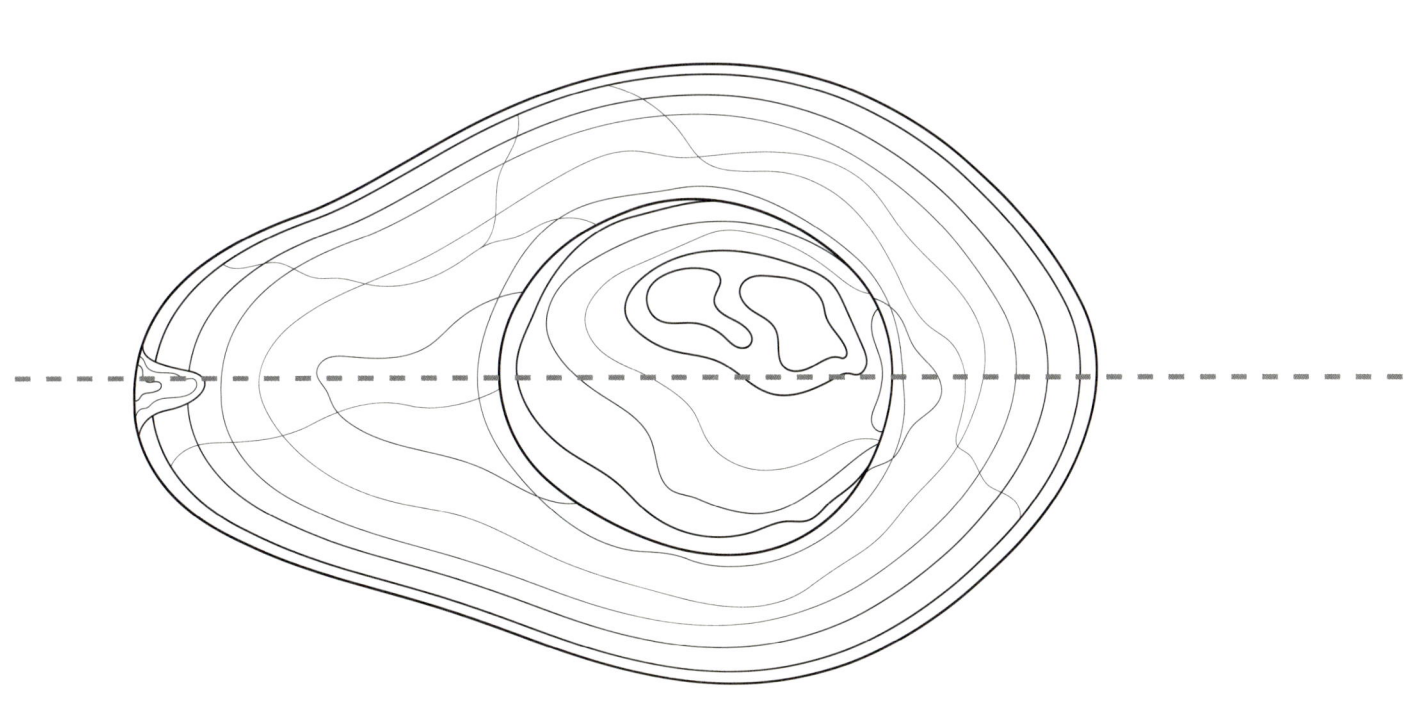

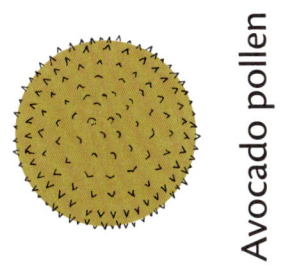

Avocado pollen

129

# GOD MADE EVERYTHING

UNIT 4 • CHAPTER 15

# ACTIVITY 43
# Greenery

Green is such a nice color! Aren't you glad God made chlorophyll green? There are so many different colors of green in the plants of the world.

Read Psalm 23:1. Shepherds look for the best pastures for their sheep to lie down in. They look for green plants for their sheep to eat and calm water for them to safely drink. The Lord is our shepherd. He takes care of us just like shepherds take care of their sheep.

Let's make a chlorophyll painting.

1. Gather some cotton swabs, an adult with sturdy scissors, some green leaves (spinach works well but waxy leaves do not), and glue.

2. Roll up different kinds of green leaves. Hold them between your fingers and rub them on the coloring page scene to paint with their color.

3. Ask an adult to cut off the ends of several cotton swabs. Glue the fuzzy ends to the green pasture to look like faraway sheep.

4. Use crayons or colored pencils to color the rest of the picture.

GOD MADE EVERYTHING

- Sing the hymn for this unit.

- Repeat the memory verse for this unit.

- Thank God for chlorophyll. Praise Him for beautiful greenery!

ACTIVITY 43: GREENERY

133

# GOD MADE EVERYTHING

UNIT 4 • CHAPTER 15

# ACTIVITY 44
# Trading Gasses

God is so wise! He makes plants take the gas we don't need and give us the gas we do need. And we take the gas plants don't need and give them the gas they do need!

In this picture, Puff is the air that carries gasses. She carries oxygen and carbon dioxide between plants and animals. Write "Oxygen" and "Carbon Dioxide" on the correct direction that Puff is blowing.

Sing the hymn for this unit.

GOD MADE EVERYTHING

Repeat the memory verse for this unit.

Thank God for the wonderful way plants and animals and people help each other breathe!

Stomata stay open all day to help the green machines make food while it is light.

UNIT 4 • CHAPTER 15

ACTIVITY 45
# Colored Leaves

Look at the nature collection you made in Unit 4. Even though your leaves are dry, they should still show what color they were.

1. Were any of them green? _____

2. Were any of your leaves yellow? _____

3. Orange? _____

4. Red? _____

5. Brown? _____

On the leaf coloring page, solve the math problems and color each leaf according to the color key.

2 = Green    4 = Orange    6 = Brown
3 = Yellow   5 = Red

GOD MADE EVERYTHING

Sing the hymn for this unit.

Repeat the memory verse for this unit.

Praise God for the beautiful colors of leaves!

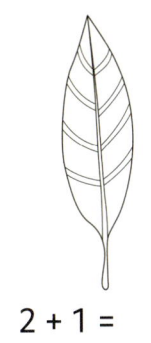

ACTIVITY 45: COLORED LEAVES

2 + 1 =

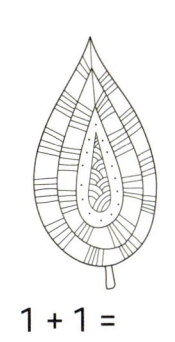

1 + 1 =

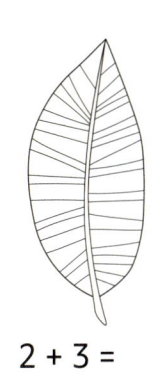

2 + 3 =

0 + 2 =

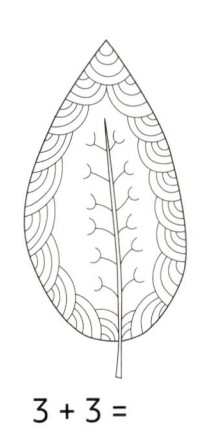

3 + 3 =

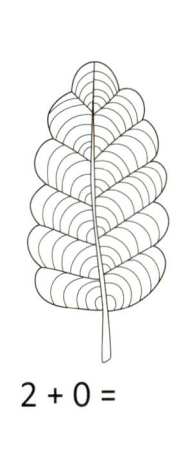

2 + 0 =

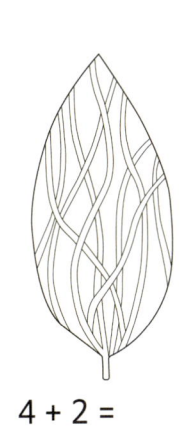

4 + 2 =

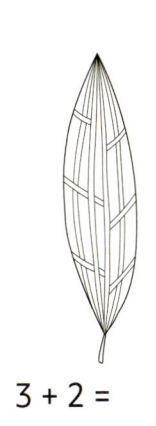

3 + 2 =

2 + 2 =

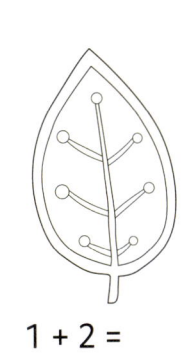

1 + 2 =

3 + 0 =

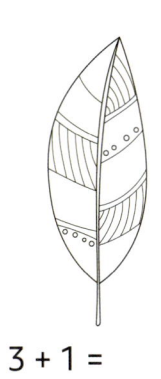

3 + 1 =

4 + 1 =

4 + 0 =

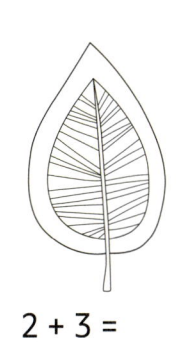

2 + 3 =

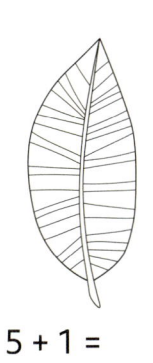

5 + 1 =

139

# GOD MADE EVERYTHING

UNIT 4 • CHAPTER 16

ACTIVITY 46
# Rhyming Practice

Poems are words written beautifully. Poems often make us think about things in a different way. Parts of the Bible are written as poems. The psalms are poems that were meant to be sung. The psalms were written long ago but we still sing them in worship today.

Poems often have words that rhyme. Rhyming words end with the same sound. "Stuck" and "truck" are rhyming words. "Pray" and "stay" also rhyme. Can you think of another set of rhyming words?

Finish each poem below with a word that rhymes with the one above it. Use the pictures to help.

I think that I shall never see
A poem as lovely as a _____.[1]

I think that I shall never push
My hand into a thorny _____.

I think that I shall spend an hour
Looking at this lovely _____.

Sing the hymn for this unit.

---
1. From the poem "Trees" by Joyce Kilmer

## GOD MADE EVERYTHING

Repeat the memory verse for this unit.

Thank God for beautiful poems. Thank Him for the psalms that we can sing to praise Him!

UNIT 4 • CHAPTER 16

ACTIVITY 47
# Ah-Choo!

There are many kinds of grasses and forbs all around the world. It is interesting to see them! It is also interesting to see their different kinds of pollen. Each kind of pollen has a different shape. We can see these shapes by looking at pictures that have been taken through a microscope.

Circle your favorite shapes of pollen in the picture.

Some people are allergic to certain kinds of pollen. It makes them sneeze and makes their eyes itch. Pollen gets in their nose as they breathe, and it tickles. Their body thinks it's a germ and tries to fight it with sneezes and

143

## GOD MADE EVERYTHING

watery eyes. Their body is doing much more than it needs to. Pollen can't hurt them, but their body thinks it can. We don't know why some people are allergic to certain pollens and other people are not.

Sing the hymn for this unit.

Repeat the memory verse for this unit.

If you are not allergic to pollen, thank God. If you are, thank Him anyway. The Bible tells us to thank Him in everything. And you can ask Him to help you with uncomfortable things. He understands and cares for you!

UNIT 4 • CHAPTER 16
ACTIVITY 48
# God Gives Us Other Plant Gifts

We have learned how important plants are for our food and fuel. We are also thankful for wood that we can build things with. There are a lot of other important uses people have found for plants:

1. Paper is made from wood.

2. Many oils come from plants.

3. Rope can be made from plant fiber.

4. Clothes can be made of cotton and linen plant fibers.

5. Some medicines and healing herbs are from plants.

6. Some paints, plastics, latex, rubber, dyes, and inks come from plants.

7. Long ago, people used leaves and corncobs for toilet paper and stringy bark for baby diapers.

8. People have invented a way to make fuel for our cars out of corn plants.

9. We all enjoy the beauty of God's creation in the plants we use in our yards, homes, and businesses.

Milky sap collected from rubber trees may one day make a rubber band.

Look around the room. Count how many kinds of things are made of plants. Write the number here: _____

Sing the hymn for this unit.

Repeat the memory verse for this unit.

Praise God for giving us plants that are so beautiful and that are useful for so many things.

UNIT 5 • CHAPTER 17
## ACTIVITY 49
# Examine Some Exoskeletons

When you look at something closely and spend time noticing things about it, you are **examining** it. Let's go find some creatures with exoskeletons to examine! Gather two or three plastic containers with lids, a cotton swab, a piece of stiff paper like a postcard, and a magnifying glass.

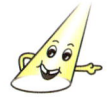

Look indoors. We don't like bugs to be inside where we live, but they often come in somehow. To catch a live bug, you can put a container over it. Then slide the stiff paper under the opening of your container before you lift it up. When you take off the paper, quickly put the lid on. Put live bugs in the refrigerator. Since crunchy creatures are cold blooded, they will slow down when they get cold but they won't die. You can examine them later.

1. Check your windowsills for dead insects. Flies have a lot of germs so use the cotton swab to gently pick them up and put them in a container. Be careful! Dried insects break easily.

2. Look in damp places like a basement or under a sink for live bugs.

3. Sometimes ants come in kitchens to find sweet or greasy things to eat. They leave a trail of smells for other ants to

GOD MADE EVERYTHING

follow. You might see a line of ants taking food back to their nest on their special trail.

4. Check the corners of rooms for spiders in their webs.

Look outside. If the weather is warm, you will be able to find more bugs than if it's cold.

1. Check on green plants and on flowers if they are blooming.

2. If it's winter and plants are dead, check the ground under dead leaves or fallen wood. Bugs can spend the winter as eggs, babies, or adults. But they always find a hiding place while they wait for warm weather.

3. If there is a pond, stream, or ocean nearby, you may be able to find creatures with exoskeletons living in the water. Watch for bubbles coming up from where they might be hiding. Look under rocks too.

Check for dead bugs in your nature collection from Chapter 4.

Use your magnifying glass to examine all the crunchy creatures you found. Look at their legs and mouth parts. Look at their interesting compound eyes. Instead of two eyes like you

## ACTIVITY 49: EXAMINE SOME EXOSKELETONS

have, most creatures with exoskeletons have a lot of eyes. All those eyes are arranged on two balls sticking out from their heads. Your eyes can move to see different directions. A bug's eyes don't move but the bug sees every direction at the same time. It does this by using information from each eye.

Sing the hymn for this unit.

Write or repeat the memory verse for this unit.

### 📖 Memory Verse

Go to the ant, you sluggard! Consider her ways and be wise, . . . [She] provides her supplies in the summer, And gathers her food in the harvest.
(Proverbs 6:6, 8)

Compound eye of a fly

149

# GOD MADE EVERYTHING

Praise God for the interesting exoskeletons you examined on the creatures you found!

UNIT 5 • CHAPTER 17

## ACTIVITY 50
# Naming

After God made all the animals, He created Adam. God brought all the animals to Adam to see what he would name them. That was a long time ago, and we don't know what Adam called all the animals. Now there are a lot of different languages in the world. Different languages have different names for things. This book is written in the English language.

The name "cricket" sounds like the summer night song that these bugs sing. That's how crickets got their English name.

Find a stiff comb. Run your thumb nail along the teeth of the comb to make a noise. God made cricket wings with tiny teeth like a comb. Try to make your comb say "cricket."

The name "grasshopper" doesn't sound like the noise grasshoppers make. It's a name that tells us what a grasshopper does.

You are called a person. But you also have a special name.

What is your name?

_____

GOD MADE EVERYTHING

On the coloring page, draw compound eyes on the grasshopper. Then color the picture. Can you think of a special name for this grasshopper?

_____

Sing the hymn for this unit.

Repeat the memory verse for this unit.

Thank God for cricket sounds and jumping grasshoppers. He designed their exoskeletons so perfectly!

Since God has named all the stars, I wonder if He has named all the grasshoppers!

ACTIVITY 50: NAMING

GOD MADE EVERYTHING

UNIT 5 • CHAPTER 17

# ACTIVITY 51
# Wise Creatures

There are four things which are little on the earth,
But they are exceedingly wise:
The ants are a people not strong,
Yet they prepare their food in the summer;
The rock badgers are a feeble folk,
Yet they make their homes in the crags;
The locusts have no king,
Yet they all advance in ranks;
The spider skillfully grasps with its hands,
And is found in kings' palaces.
(Proverbs 30:24-28)

These Bible verses tell about four creatures that are small. They are also wise. They are called wise because of the interesting and wonderful things they can do. They do these things because our wise God made them that way. You can be wise too if you do the things God made for you to do!

The hymn for this unit sings about all creatures, even the small ones (like ants, badgers, locusts, and spiders). When you sing, listen for creatures being wise and wonderful. When you sing the word "wise," hold up three fingers to make a "W" in the air.

GOD MADE EVERYTHING

Draw the three creatures with exoskeletons found in Proverbs 30:24-28.

Sing the hymn for this unit.

Repeat the memory verse for this unit.

Thank God for the wisdom He used to make wise creatures. Ask Him to help you be wise in the things you do.

UNIT 5 • CHAPTER 18

# ACTIVITY 52
# How Many?

How many body parts does an insect have? Let's see if you can remember! Draw straight lines from the words on the left to the correct numbers on the right, crossing over the butterfly's body.

Body Sections                                           4

Antennae
(This word means "more                                  6
than one antenna.")

Legs                                                    2

Wings                                                   3

157

GOD MADE EVERYTHING

Now make the lines you drew into butterfly wings by drawing edges on them and coloring them.

Sing the hymn for this unit.

Repeat the memory verse for this unit.

Praise God for interesting insects!

UNIT 5 • CHAPTER 18

ACTIVITY 53

# Big Changes!

Metamorphosis seems like a miracle! Can you imagine waking up one morning and finding that you have wings? And then imagine that you could fly with your new wings without even learning how! Maybe getting wings seems like a miracle to an insect too.

What happens next? On the Butterfly Metamorphosis coloring page, circle the butterfly eggs. Then draw an arrow to what happens next. Keep drawing arrows to show the correct order of the changes in a butterfly's life. Color the picture.

Answer these questions about metamorphosis. Choose from the words below:

    Moth        Grasshopper        Butterfly

1. Which two insects have a pupa called a chrysalis?

   _____ and

   _____

GOD MADE EVERYTHING

2. Which insect has a chrysalis and a cocoon?

   _____

3. Which insect does not ever have a pupa?

   _____

Sing the hymn for this unit.

Repeat the memory verse for this unit.

Praise God for the amazing ways metamorphosis changes baby insects into flying adults!

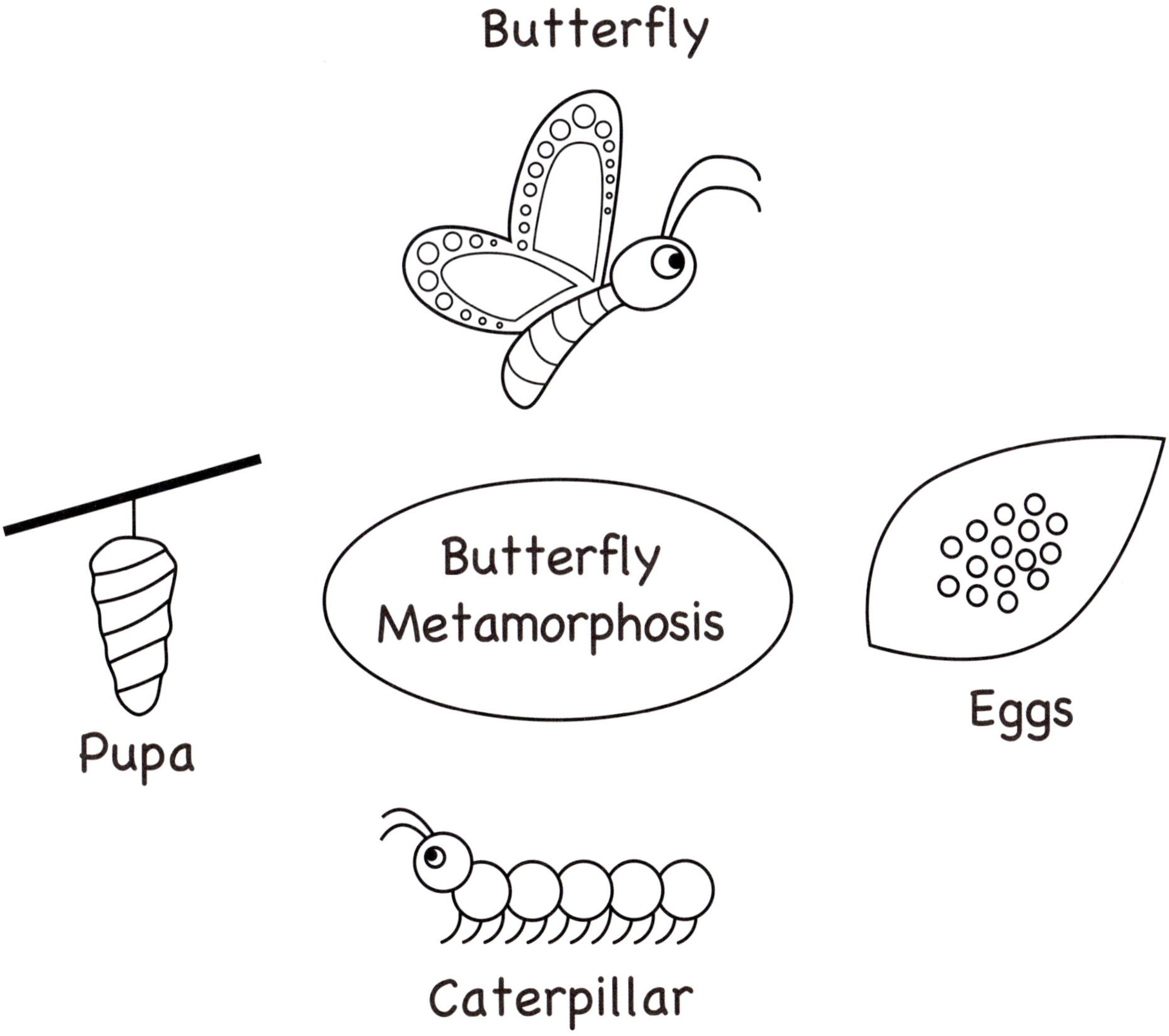

ACTIVITY 53: BIG CHANGES!

Butterfly

Butterfly Metamorphosis

Pupa

Eggs

Caterpillar

161

# GOD MADE EVERYTHING

UNIT 5 • CHAPTER 18
ACTIVITY 54
# Be a Mess-Eating Insect!

God made insects that eat rotting plants or animals. These insects help turn the rotting plants and animals into dirt. Once these messes are part of the dirt, we don't notice them anymore. They are cleaned up! The dirt now has good things in it that help new plants grow. Thank you, mess-eating insects!

Pretend you are a mess-eating insect. Clean up your room or another place your parents suggest. As you pick things up, pretend you are eating them. When you put the things where they belong, pretend they are going into the dirt. If you do a good job, you won't notice the mess anymore.

Sing the hymn for this unit.

When rhinoceros beetles are larvae, they eat rotting plants. When they become adults like this one, they eat nectar, sap, and fruit.

# GOD MADE EVERYTHING

Repeat the memory verse for this unit.

Thank God for mess-eating insects that keep the ground clean and make good things for plants to use.

UNIT 5 • CHAPTER 19

## ACTIVITY 55
# Lots of Little Locusts

Grasshoppers love to eat plants. Usually they don't eat enough plants to make a difference in a wild field. But sometimes there isn't enough rain for plants to grow well. Then, if a locust swarm comes, they might eat everything growing in the field!

In the Bible, God uses grasshoppers to teach us many things:

- God talks about His power and our weakness. He says we are like little grasshoppers. (Isaiah 40:22)

- When God wants to show that there are a lot of people coming, He talks about them coming as locusts. (Judges 7:12)

- Often grasshoppers are called locusts in the Bible because they destroy crops. Read Exodus 10:12-19. What color was missing in Egypt right before God blew the locusts into the sea?

_____

- Sing the hymn for this unit.

Locust swarm in Africa. These people are trying to catch the locusts for food.

- Repeat the memory verse for this unit.

- Thank God that He loves us even though we are like weak little grasshoppers. Ask Him to protect crops in Africa from locusts.

UNIT 5 • CHAPTER 19
# ACTIVITY 56
# Bee Maze

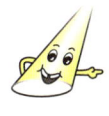

Help the bee find its way to the hive!

Sing the hymn for this unit.

GOD MADE EVERYTHING

Repeat the memory verse for this unit.

Praise God for wasps that eat the caterpillars that eat our garden plants. Thank Him for sweet honey!

UNIT 5 • CHAPTER 19

## ACTIVITY 57
# What Am I?

Let's see what you remember about the insects we have learned about. Write the first letter of the insect's name behind the clue. Use these insect names:

D = Dragonfly     G = Grasshopper     F = Fly
T = True Bug      B = Beetle          M = Moth
P = Praying Mantis   W = Wasp

1. I have big clear wings that I keep folded and hidden when I'm not flying. What am I? _____

2. When I was a baby, I had gills like a fish. What am I? _____

3. I have tiny scales on my wings. What am I? _____

4. I have special knobs instead of a second set of wings. What am I? _____

5. I drink my food by poking my mouth parts in it. What am I? _____

6. When I was a baby, I grew up inside a caterpillar. What am I? _____

GOD MADE EVERYTHING

7. My back legs are bigger than my other four legs. What am I? _____

8. My front legs are bigger than my other four legs. What am I? _____

Sing the hymn for this unit.

Repeat the memory verse for this unit.

Praise God for so many interesting kinds of insects!

UNIT 5 • CHAPTER 20
ACTIVITY 58
# Air in Water

Oxygen is a gas that animals need in order to live. But how does oxygen get in water for underwater animals to use? God has made pieces of oxygen so tiny that they can mix with water and stay there. This is called **dissolved oxygen**.

Underwater creatures could use up all the dissolved oxygen in water. But God has made a way to put oxygen back in. Oxygen can travel from the air into the water at its surface. Oxygen can also be dissolved into water when waves or streams splash. The water drops pick up oxygen as they fly through the air. People help put oxygen into water too. They put fountains in fish ponds to make more oxygen for the fish.

A waterfall gives enough dissolved oxygen for the koi fish in this pond.

# GOD MADE EVERYTHING

Let's see if we can observe oxygen in water.

## Oxygen In Water!

**You will need:**

- A large frying pan
- Water
- Stove
- An adult's help

**Instructions:**

1. Put 1 to 1 1/2 inches (2-4 cm) of water in the frying pan.
2. Heat the water over low heat.
3. Watch for tiny bubbles to form on the bottom of the pan. The bubbles are made of oxygen that couldn't stay dissolved when the water got too warm.
4. If the water gets hotter, the tiny bubbles will rise and pop in the air.
5. If the water keeps getting hotter, it could boil. The bubbles from boiling water are not oxygen. They are water vapor.

Sing the hymn for this unit.

Repeat the memory verse for this unit.

Praise God that oxygen can float around in air and in water so that His creatures have what they need to breathe!

UNIT 5 • CHAPTER 20

ACTIVITY 59

# Make a Web!

God has made a way for each creature to get its food. Very few animals build traps to catch food. But spiders do! They trap the insects they eat in a web made from their own bodies.

Let's see if you can make a strong web!

## Make Your Own Web

**You will need:**

- Three hard backed chairs with back posts
- A spool of thread
- A pair of scissors

**Instructions:**

1. Arrange the chairs with their seats facing out and their backs together to make three sides of a square.
2. Stand at the open side of the square.
3. Tie the end of your thread (spiderweb) to a chair post.
4. Stretch the thread across the open space and loop it around another chair's back post.
5. Keep stretching your thread across the square at different angles to different chair posts.

GOD MADE EVERYTHING

6. When you have a lot of crisscrosses, you can weave your thread over and under the other threads to make your web stronger.
7. When the web is strong, cut the thread and tie it to a post.
8. Test the strength of your web by tossing "insects" into it. These could be stuffed animals, hats, socks, or mittens.

Sing the hymn for this unit.

Repeat the memory verse for this unit.

Praise God for spiders. They are amazing weavers!

UNIT 5 • CHAPTER 20

ACTIVITY 60
# Butterfly Scales

Butterfly and moth wings are covered with tiny flat plates called scales.

To make this butterfly art project, you will need: sequins for scales and glue.

Using the butterfly outline, put glue where you would like the sequins to be. Work in one small area at a time so the glue doesn't dry before you put sequins on it. Cover the butterfly wings with your beautiful sequins!

Some people make interesting art projects with butterfly wing scales. They have to use a microscope to make the picture. We would have to use a microscope to see their artwork! But an adult could find an internet picture for you by searching for "butterfly scale art."

Sing the hymn for this unit.

Repeat the memory verse for this unit.

Praise God for tiny shiny scales that make such beautiful butterflies!

175

GOD MADE EVERYTHING

ACTIVITY 60: BUTTERFLY SCALES

177

GOD MADE EVERYTHING

UNIT 6 • CHAPTER 21

# ACTIVITY 61
# Swim Bladders

Bony fish have something called a **swim bladder**. Swim bladders are pouches inside fish that can fill with air. The air helps the fish stay at a certain level in the water. The swim bladder helps a fish save energy because the fish doesn't have to use its fins to hold itself up or down.

Let's see how a swim bladder works!

Inflated swim bladder of Kokanee salmon

## Swim Bladder

**You will need:**

- Two uninflated balloons
- Two of the same kind of coin
- A waterproof marker
- A bucket of water

**Instructions:**

1. Using the waterproof marker, draw fish eyes, a mouth, and fins on each uninflated balloon.
2. Put a coin in each balloon.
3. Blow air into one of the balloons. Leave the other balloon uninflated.
4. Tie the balloons closed.
5. Put the balloons in the bucket of water. Do you see how the amount

GOD MADE EVERYTHING

of air in a fish's swim bladder might help it stay at a certain level in the water?

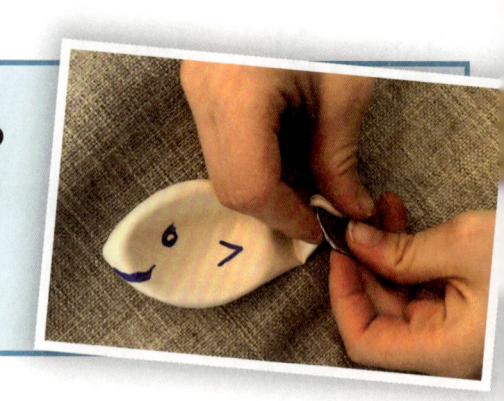

_____

Answer these questions for a fish:

1. It's winter. Ice is covering the top of the lake. The water is warmer at the bottom of the lake. Should the fish make its swim bladder full or empty?

   _____

2. It's summer. Mosquitoes are landing on the lake. The fish wants to stay near the top of the water and watch for a bug to eat. Should its swim bladder be full or empty?

   _____

Sing the hymn for this unit.

Write or repeat the memory verse for this unit.

ACTIVITY 61: SWIM BLADDERS

## 📖 Memory Verse

In God's hand is the life of every living thing and the breath of all mankind. (Job 12:10)

Thank God for the interesting little balloons He put inside fish!

UNIT 6 • CHAPTER 21

ACTIVITY 62

# You Have Something Sharks Have!

Sharks and rays have skeletons made of cartilage instead of bone. You have cartilage too!

- Feel your ears. Try bending them and moving them around. They are made of cartilage inside! Cartilage gives them their shape.

- Your nose is made of cartilage too. Try moving it around.

- Feel your throat. That stiff, bendy tube in front is where air travels when you breathe. It is also made of cartilage.

Sharks do have some things that are as hard as bones: their teeth! They have an extra row of teeth behind their front teeth to replace the ones that are always falling out. Your teeth have long roots, so they don't fall out as easily as shark teeth with their short roots.

- Sing the hymn for this unit.

### ACTIVITY 62: YOU HAVE SOMETHING SHARKS HAVE!

Repeat the memory verse for this unit.

Thank God that your teeth don't keep falling out like shark's teeth do.

UNIT 6 • CHAPTER 21

# ACTIVITY 63
# Precious Pearls

God's people are part of His kingdom. Jesus says His kingdom is very valuable:

> "The kingdom of heaven is like a merchant seeking beautiful pearls, who, when he had found one pearl of great price, went and sold all he had and bought it." (Matthew 13:45-46)

Pearls are beautiful creations of God. They are a hidden treasure that people try to find. People make special jewelry from pearls. God does not make pearls underground where jewels are found. He makes them in oysters!

Pearls are made in the ocean, between the two matching shells of oysters. A piece of sand gets inside the shells. It bothers the oyster like a rock in your shoe would bother you. The oyster can't take off its shell to get the sand out. Instead, its body makes a lovely white coating for the sand. The coating makes the sand smooth and round. It gets bigger and bigger, but it doesn't bother the oyster anymore. It takes about seven years for an oyster to make a pearl from a piece of sand.

Color the picture of the octopus with a pearl. Leave the pearl white.

ACTIVITY 63: PRECIOUS PEARLS

GOD MADE EVERYTHING

ACTIVITY 63: PRECIOUS PEARLS

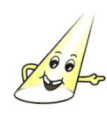

Sing the hymn for this unit.

Repeat the memory verse for this unit.

Lord, You are our treasure! Thank You for making beautiful pearls!

UNIT 6 • CHAPTER 22

ACTIVITY 64

# Grow into a Frog

Let's act out the stages of a frog's life cycle!

## Frog Life Cycle

**You will need:**

- An adult t-shirt
- A pillowcase
- A sturdy rubber band

**Instructions:**

1. Bring your arms and legs inside the shirt.
2. Have an adult gather the shirt's hem into a sturdy rubber band.
3. Bring your head inside the shirt. You are a frog egg!
4. Wiggle around inside the shirt and pretend to be growing. You are an embryo!
5. Bring out your head. You are hatching! Now let's pretend the big shirt is no longer your egg. Pretend it's now your skin.
6. Ask an adult to tuck a corner of the pillowcase into the rubber band to be a tail. While lying on your tummy, ask a helper to move your tail from side to side. You are a swimming tadpole! Have an adult take off the rubber band. Pop your legs out of the opening. Ask the adult to attach your tail to the back hem of the shirt. You are a tadpole with back legs!

ACTIVITY 64: GROW INTO A FROG

7. Pop your arms out of the shirt through its arm holes. You are a tadpole with four legs!
8. Have the adult take off your tail. It could be rolled up first to show that it is getting smaller. Pretend to crawl out of the water. You are a frog!
9. Use your long legs to jump around on land. Then jump back into the water to get your skin wet again or to escape from an enemy.

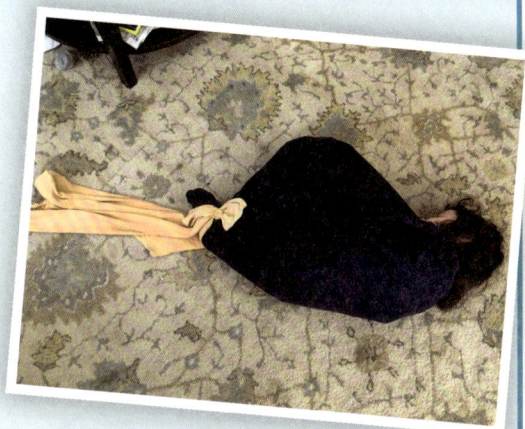

Sing the hymn for this unit.

Repeat the memory verse for this unit.

Thank God for the interesting metamorphosis of amphibians!

UNIT 6 • CHAPTER 22

ACTIVITY 65
# Webbed Feet

Let's see how webbing helps frogs swim!

## How Do Frogs Swim?

**You will need:**

- A plastic bag big enough for your fingers to spread out inside of it

**Instructions:**

1. Put water in a sink or bathtub. Make it as deep as the width of your hand with your fingers spread out.
2. Spread out your fingers. Move your hand through the water. Can you feel your fingers pushing against the water?
3. Now put the plastic bag over your hand. Spread out your fingers, and move your hand through the water. Can you push more water with the plastic bag over your hand?
4. Do you see how webbed feet help a frog swim? Webbed feet can push more water so the frog can swim faster.

Sing the hymn for this unit.

- Repeat the memory verse for this unit.

- Praise God for His wonderful work in making special feet for frogs!

UNIT 6 • CHAPTER 22

# ACTIVITY 66
# Salamander Coloring Page

Color the salamander. If you want it to have poisonous skin, color it with bright colors. God gave poisonous salamanders bright skin. Their skin is a warning to tell other animals not to eat them.

Always wash your hands after touching or holding amphibians. They might have germs or poison on their skin.

Sing the hymn for this unit.

Repeat the memory verse for this unit.

Praise God for making an animal that can grow new body parts when it needs to!

ACTIVITY 66: SALAMANDER COLORING PAGE

193

# GOD MADE EVERYTHING

UNIT 6 • CHAPTER 23

ACTIVITY 67

# Find the Warmth

Circle the animal name below that rhymes with "Reptile."

Alligator    Snake    Lizard    Crocodile    Tortoise    Turtle

Reptiles look for warm places in the chilly morning. Go outside and look for sunny places. Put your hand on them to feel how warm they are. Now find some shady places. Put your hand on each one to feel how cool it is. What was the warmest place you found?

_____

_____

# GOD MADE EVERYTHING

If a reptile gets too hot, it must move to a cooler place. Describe the coolest place you found.

_____

_____

Sing the hymn for this unit.

Repeat the memory verse for this unit.

Praise God for reptiles! They need to search for God's sunlight and warmth.

UNIT 6 • CHAPTER 23

ACTIVITY 68

# Sense Like a Snake

Let's pretend to sense things the way a snake does! You will need a helper.

1. Close your eyes almost all the way so things look blurry.

2. At the same time, flick your tongue in and out. Pretend that you can smell lots of things with your tongue.

3. Put your hands over your ears so you can't hear well.

4. Have a helper put warm hands on your cheeks. Pretend you are feeling their warmth through the snake pits on the sides of your head.

You are probably wondering how a snake knows anything with such strange senses. Remember that God does all things very well. He made the snake's senses work just right for its life!

GOD MADE EVERYTHING

Circle ten differences between the two chameleon pictures.

Sing the hymn for this unit.

Repeat the memory verse for this unit.

Praise God that your senses work just right for you and that a snake's senses work just right for it!

198

UNIT 6 • CHAPTER 23

ACTIVITY 69
# Tortoise or Turtle?

Circle the creature that is not a tortoise.

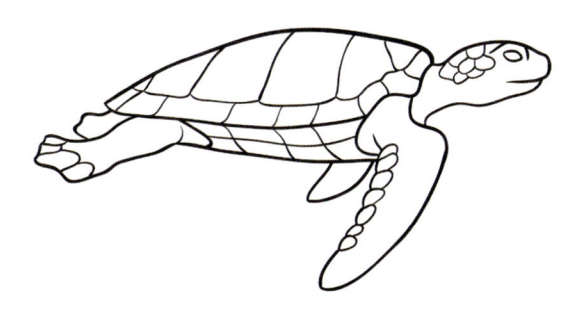

# GOD MADE EVERYTHING

Sing the hymn for this unit.

Repeat the memory verse for this unit.

Praise God for so many kinds of reptiles and all their differences!

UNIT 6 • CHAPTER 24
## ACTIVITY 70
# Make Print Fossils

Many fossils are animal bones. Other fossils are footprints or prints of an animal's skin. Some fossils are even prints of leaves.

Let's see how print fossils might have been made in soft mud!

1. With an adults help, make craft dough (recipe included). You could also use purchased modeling dough.

2. Gather some things like sea shells, sticks, leaves, and plastic animal toys that won't get ruined in the dough.

3. Make print fossils by pressing these things into the dough and lifting them out.

4. Save the dough in a plastic bag at room temperature for the next activity.

Sing the hymn for this unit.

Repeat the memory verse for this unit.

GOD MADE EVERYTHING

## Homemade Craft Dough

**You will need:**

- 1 cup (125g) flour
- 1/4 cup (75g) salt
- 1/2 cup plus one tablespoon (135ml) water
- 3 tablespoons (45ml) lemon juice
- 1 tablespoon (15ml) cooking oil

**Instructions:**

- Mix the flour and salt in a mixing bowl.
- In a small sauce pan, combine the water and lemon juice. Bring the mixture to a boil. Remove from heat and add the liquid to the flour mixture.
- Mix with a spoon.
- Stir in oil, kneading with your hands if necessary.
- Allow the dough to cool. When cool, add more flour if the dough is too sticky.

Thank God that we can learn about dinosaurs from fossils!

UNIT 6 • CHAPTER 24

ACTIVITY 71
# Behemoth Sculpture

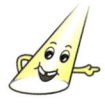

Make a sculpture of how the behemoth in the Bible may have looked. Use your craft dough saved from Activity 70. The pictures below show the steps to make a flat sculpture using about a fourth of the recipe's dough.

1. On a piece of foil, divide your dough into three parts. Set one part aside. This will be the body.

2. Divide one of the other parts in half. Divide the last part into four pieces.

GOD MADE EVERYTHING

3. Roll one of the half pieces into a neck and the other half into a tail. Make the four small pieces into legs.

4. Attach the legs, neck, and tail to the body.
5. Set your sculpture in a dry place. Let it sit for several days. After it has hardened, you can paint it if you like.

Job 40:15-24 says behemoth:

- Was a plant eater

- Had strong hips, thighs, and stomach

- Moved its tail like a cedar tree

- Had big bones as strong as metal

- Wasn't bothered when the river rushed into its mouth and eyes

ACTIVITY 71: BEHEMOTH SCULPTURE

- Sing the hymn for this unit.

- Repeat the memory verse for this unit.

- Praise God for such a huge animal that shows His power!

UNIT 6 • CHAPTER 24
## ACTIVITY 72
# Draw Leviathan

Job 41 is an exciting chapter to read! It talks about Leviathan, a huge scary creature that could swim in the ocean and come onto land. The Bible says Leviathan:

- Was too fierce for people to control

- Had terrible teeth

- Had scales joined tightly together

- Breathed out fire

- Could raise itself up

- Crashed around on land and churned up the ocean

- Had sharp scales on its stomach

An artist named Gustave Dore drew this picture of what he thought Leviathan might have looked like.

206

## ACTIVITY 72: DRAW LEVIATHAN

Maybe Leviathan had sharp scales like these small lizards do.

- Using paper and crayons, colored pencils, or markers, draw what you think Leviathan may have looked like.[1]

- Sing the hymn for this unit.

- Repeat the memory verse for this unit.

- Praise God for such a fierce creature that shows His amazing power!

---

1. Here's an interesting article on Leviathan: https://answersingenesis.org/dinosaurs/drawing-out-biblical-leviathan/

UNIT 7 • CHAPTER 25

ACTIVITY 73
# Wings Push Air

- Look at these flying pigeons. Can you find the flat flight feathers on their tails and on the back edges of their wings?

- Color the picture. Make the bird's flight feathers a different color than the other feathers on its body.

- Sing the hymn for this unit.

- Write or repeat the memory verse for this unit.

ACTIVITY 73: WINGS PUSH AIR

# GOD MADE EVERYTHING

ACTIVITY 73: WINGS PUSH AIR

### Memory Verse

But those who wait on the LORD
Shall renew their strength;
They shall mount up with wings like eagles.

(Isaiah 40:31)

Thank God for the way a bird's feathers help it push air!

UNIT 7 • CHAPTER 25

ACTIVITY 74
# Learn to Whistle!

Springtime is filled with the lovely singing of birds! Birdsong is different from any other creature's voice. People talk and sing with sound from their throats. A bird is the only animal that makes its sound in its chest.

Birdsongs come from the place where a bird's two lungs start. That's why birds can sometimes make two sounds at the same time! God has given some birds twelve tiny muscles to make their lovely songs.

Some people can whistle like birds. God made people different from birds. We have to whistle with our lips.

Let's see if you can learn to whistle! Here's how:

1. Wet your lips and pucker them.

2. Blow air through your lips, softly at first. You should hear a tone.

3. Blow harder, keeping your tongue relaxed.

4. Move your lips, jaw, and tongue to create different tones.

Sing the hymn for this unit.

Repeat the memory verse for this unit.

Praise God for beautiful birdsong.

213

UNIT 7 • CHAPTER 25
## ACTIVITY 75
# Bird Food

How does food travel inside a bird after the bird eats? With a pencil, trace the path of food in the bird.

214

ACTIVITY 75: BIRD FOOD

Use words from the clue box to fill in the blanks below:

Stomach        Crop        Intestine        Gizzard

1. The _____ grinds food. Sometimes it uses small stones to help.

2. The _____ adds juices to the food to help soften it.

3. The _____ takes vitamins and energy out of the food and puts it into the bird's blood.

4. The _____ stores food that the bird has eaten until the stomach is ready.

Sing the hymn for this unit.

Repeat the memory verse for this unit.

Praise God that a bird's food is used in a way that helps it fly!

UNIT 7 • CHAPTER 26
ACTIVITY 76

# Beaks Are Tools

- Let's see what it might be like to use a bird beak as a tool.

- Gather some popped popcorn, a small bowl for a nest, an adult with sturdy scissors, and a cardboard egg carton with tall, pointed dividers down the middle.

- Ask the adult to cut out a set of two pointed dividers from the egg carton. Keep them joined together. This will be a bird beak.

- Put your thumb and a finger into the beak. Pick up popcorn with the beak and put it into the nest to feed your chicks.

- Sing the hymn for this unit.

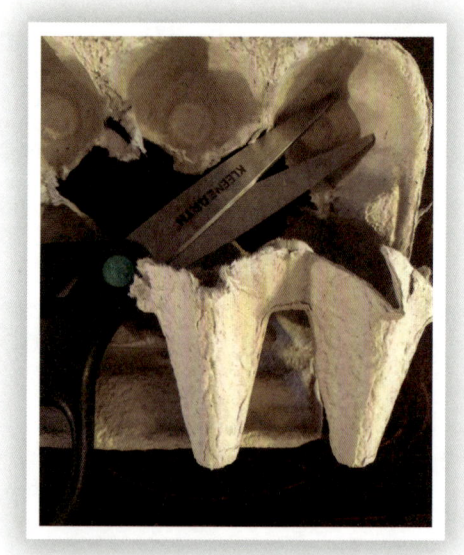

ACTIVITY 76: BEAKS ARE TOOLS

Repeat the memory verse for this unit.

Praise God for the gift of beaks that He gave birds to use as tools!

UNIT 7 • CHAPTER 26

## ACTIVITY 77
# Bird Feet

What kind of feet does this bird have? Circle the correct picture.

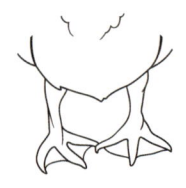

Let's see how perching works! You will need a broomstick and two stacks of books about six inches (15 cm) tall.

1. Place the two stacks of books about three feet (1 m) apart.

2. Put the broomstick across the books to make a perch.

3. Kneel in front of the stick and spread out your hands like bird feet.

218

ACTIVITY 77: BIRD FEET

4. Pretend your hands are a bird's feet landing on the stick. Your thumb should be on one side of the stick, and your fingers should be on the other side.

5. Push down with your weight on your hands.

6. Pretend to land on the stick again, but try to keep your hands open while you lean on them. Did your hands want to close? _____

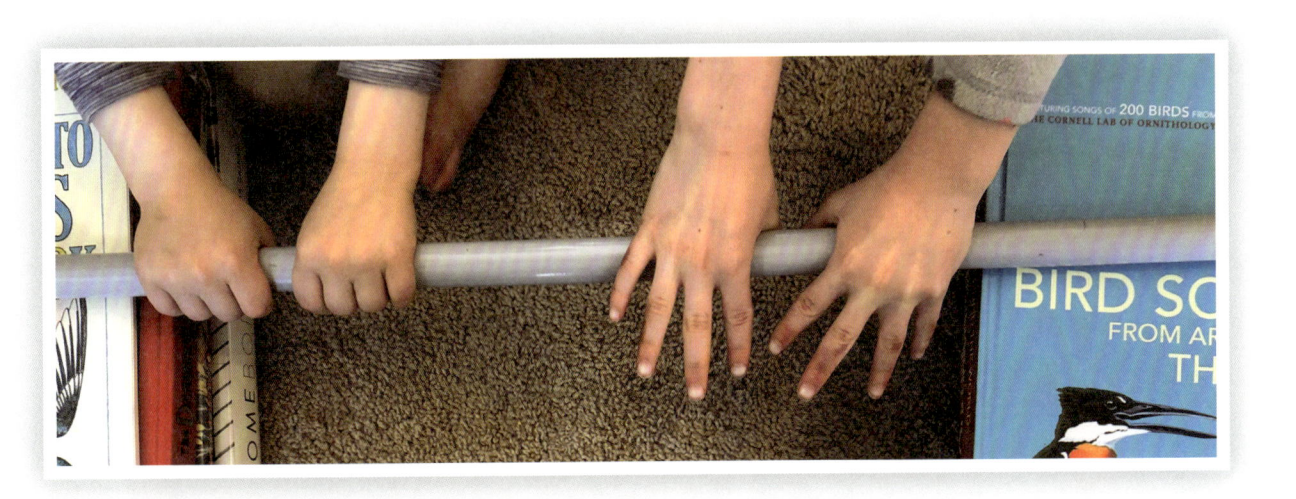

Inside your hands, you have stiff tendons that attach your muscles to your bones. When you leaned on the stick, you were pushing on these tendons. This made your tendons pull on the muscles to close your hands a little.

Something like this happens when birds perch. The weight of their body makes their toes automatically curl around the branch. But God made bird toes curl so tightly that the bird can even sleep without falling off!

219

GOD MADE EVERYTHING

Sing the hymn for this unit.

Repeat the memory verse for this unit.

Praise God for keeping birds safely on their branches while they sleep! Thank Him for keeping you safe while you sleep.

UNIT 7 • CHAPTER 26

ACTIVITY 78
# Eggs

Let's learn about an egg and then use it for something yummy! You will need a small bowl, a small plate, and all the ingredients in the cookie recipe.

Look at the outside of the egg. You are looking at its **shell**. The shell keeps the chick from being squished when its parents sit on it. The shell has tiny holes where oxygen comes in and carbon dioxide goes out.

Not all eggs have chicks growing in them. The egg you are holding doesn't have a chick inside it. Crack the egg by tapping it on a small plate until it has a crackly dent. Put your thumbs in the dent. Pull the shell apart over the bowl to let the liquid fall out. If a piece of shell also falls into the bowl, you can scoop it out with one of the egg shells in your hand.

The yellow part inside the egg is called a **yolk**. The yolk is food for the chick.

The clear part of the egg is called the **white**. The egg white is like a pillow all around the chick. It keeps the chick from getting bumped against the shell. The egg white also kills germs and gives water to the chick.

# GOD MADE EVERYTHING

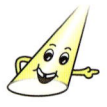

Make bird nest cookies using the egg you cracked.

Sing the hymn for this unit.

Repeat the memory verse for this unit.

## Bird Nest Cookies

**You will need:**

- 3/4 cup (170 g) butter, at room temperature
- 1/4 cup plus 2 tablespoons (75 g) sugar
- 1/2 teaspoon (2.5 ml) pure vanilla extract
- 1 3/4 cups (240 g) all-purpose flour
- 1/2 teaspoon (3 g) salt
- 1 egg, beaten (from activity above)
- 1/2 cup (50 g) shredded coconut
- 5 ounces (150 g) jelly beans

**Instructions:**

1. With an electric mixer, cream together the butter and sugar. Add the vanilla and mix. Separately, mix together the flour and salt. With the mixer on low speed, add the flour mixture to the creamed butter and sugar. Mix. Wrap the dough in plastic and chill for 30 minutes.
2. Preheat the oven to 350 degrees F (150 C).
3. Roll the dough into 1 1/4 inch (3 cm) balls. Dip each ball into the beaten egg and then roll it in the coconut. Place the balls on an

ACTIVITY 78: EGGS

ungreased cookie sheet. Press a light dent into the top of each cookie with your thumb. Bake for 20 to 25 minutes, until the coconut is a golden brown. Cool and put two or three jellybean "eggs" in each "nest."

Praise and thank God for all the special parts of an egg!

UNIT 7 • CHAPTER 27

ACTIVITY 79

# Help Migrating Birds

Migrating birds are often very tired after flying to their next home. Sometimes they die if they can't find food quickly after flying across an ocean. You can help migrating birds in the spring and fall by putting out bird feeders with birdseed, suet, fruit, and hummingbird food. You can also feed the birds that spend the winter near your home.

- Color the picture.

- Sing the hymn for this unit.

- Repeat the memory verse for this unit.

- Thank God that He feeds the birds. Ask Him to protect them from hunger.

ACTIVITY 79: HELP MIGRATING BIRDS

225

GOD MADE EVERYTHING

UNIT 7 • CHAPTER 27

## ACTIVITY 80
# Finding North

God has given birds something inside them that helps them know which directions are north and south. Birds use this special gift when they migrate. People have invented a tool called a **compass** to help us find the same directions.

- For this activity you will need a compass and some sidewalk chalk or a stick for marking the ground. Go outside where you can walk around.

- Look at your compass. Do you see the **N** for **north** and the **S** for **south**? Notice the **needle** that can spin. It always points north.

- Take a walk and look at your compass now and then. Draw arrows on the ground to point north.

- Sing the hymn for this unit.

- Repeat the memory verse for this unit.

- Praise God for the special ways birds and people can tell direction on Earth!

UNIT 7 • CHAPTER 27
ACTIVITY 81
# Nesting

In this verse, God is showing that He is the only one who makes eagles fly high to build nests:

"Does the eagle mount up at your command,
And make its nest on high?"
(Job 39:27)

- On the next page, draw a line to show the right way to the eagle's nest.

- Sing the hymn for this unit.

- Repeat the memory verse for this unit.

- Praise God that birds automatically know how to make nests for their babies in just the right places!

ACTIVITY 81: NESTING

229

GOD MADE EVERYTHING

UNIT 7 • CHAPTER 28

# ACTIVITY 82
# Oil and Water Don't Mix

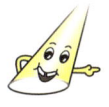

Let's see how a dipper's oil helps its feathers stay dry!

Water running off a dipper's feathers

1. You will need a helper, some vegetable oil, and a container of water that your hand will fit into.

2. Ask your helper to cover one of your hands in oil.

231

GOD MADE EVERYTHING

3. Quickly dip each hand in and out of the water. Watch the water on your hands. Do you see a difference in how the water acts on each hand? What do you notice?

_____

_____

Sing the hymn for this unit.

Repeat the memory verse for this unit.

Thank God for the way oil helps dippers stay dry and comfortable!

UNIT 7 • CHAPTER 28
ACTIVITY 83
# Birds Named for Their Songs

Each kind of bird has its own interesting song. People have named some birds by the sound of their songs.

Ask an adult to help you listen to these birdsongs on the internet. How do you think these birds got their names?

- Hummingbird
- Killdeer
- Chickadee
- Whip-poor-will
- Cuckoo

Cuckoo singing

Sing the hymn for this unit.

Repeat the memory verse for this unit.

Thank God for creating so many different birdsongs!

UNIT 7 • CHAPTER 28

## ACTIVITY 84
# Be a Honeyguide!

To be a special kind of honeyguide you will need a bottle of honey and some helpers that you can guide.

1. Write this Bible verse about honey on a piece of paper:

   How sweet are Your words to my taste,
   Sweeter than honey to my mouth!
   (Psalm 119:103)

2. Hide the Bible verse and honey together somewhere outside.

3. Pretend to be a honeyguide. Chatter at your helpers to gradually lead them to the honey.

4. Whose words are sweeter than honey?

   _____

Sing the hymn for this unit.

ACTIVITY 84: BE A HONEYGUIDE!

- Repeat the memory verse for this unit.

- Thank God for sweet honey and His good words in the Bible!

# GOD MADE EVERYTHING

UNIT 8 • CHAPTER 29
ACTIVITY 85
# Mammals Feed Their Babies

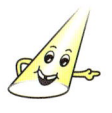

Color the picture of a cow nursing her calf.

GOD MADE EVERYTHING

## ACTIVITY 85: MAMMALS FEED THEIR BABIES

Sing the hymn for this unit.

Write or repeat the memory verse for this unit. Instead of the word "thousand," you may write the number 1,000.

> ### Memory Verse
>
> "For every beast of the forest is Mine,
> And the cattle on a thousand hills.
> I know all the birds of the mountains,
> And the wild beasts of the field are Mine."
>
> (Psalm 50:10-11)

# GOD MADE EVERYTHING

Praise God for His gift of milk for baby animals. Thank Him that cows and goats make enough milk for people to have some too!

UNIT 8 • CHAPTER 29

# ACTIVITY 86
# Quills for Protection

Let's make an edible porcupine craft!

## Edible Porcupine

**You will need:**

- A ripe pear or kiwi fruit
- A box of toothpicks
- 11 raisins
- A plate

**Instructions:**

1. Wash and dry the fruit and set it on the plate.
2. If you're using a pear, remove its stem. The stem end will be the porcupine's nose. If you're using a kiwi, choose one end of the kiwi to be the nose.
3. At the other end of the piece of fruit, poke in several whole toothpicks to make a long tail.
4. Break some toothpicks in half. These will be quills.
5. Poke the toothpick halves into the fruit, starting at the tail and working toward the head. Push the broken end of each toothpick

Porcupine quills are hard, sharp hairs they use for protection.

241

# GOD MADE EVERYTHING

> into the fruit. The quills should be angled backward like they are on a real porcupine.
> 6. Leave the head area bare.
> 7. Using three toothpick halves, attach raisins for eyes and a nose.
> 8. For each leg, use half a toothpick and two raisins.

Sing the hymn for this unit.

Repeat the memory verse for this unit.

Praise God for His interesting way of protecting porcupines!

UNIT 8 • CHAPTER 29

ACTIVITY 87
# Animal Baby Names

Here is a list of animal babies. Choose the correct name for each animal's baby and write it in the blanks. Some letters have been written for you.

Calf        Fawn        Joey        Kitten      Puppy
Cub         Foal        Kit         Pup

1. Kangaroo:   J __ __ __

2. Dog:        P __ __ __ __

3. Horse:      F __ __ l

4. Seal:       P __ __

5. Whale:      C __ __ __

6. Fox:        K __ __

7. Cat:        K __ __ __ __ __

8. Bear:       C __ __

9. Deer:       F __ __ n

GOD MADE EVERYTHING

Sing the hymn for this unit.

Repeat the memory verse for this unit.

Praise God for cute little mammal babies!

UNIT 8 • CHAPTER 30
# ACTIVITY 88
# Special Heads and Feet

Draw lines to match the mammal's special gifts with the correct picture.

Tusks

Kudu

Horns

Reindeer

Antlers

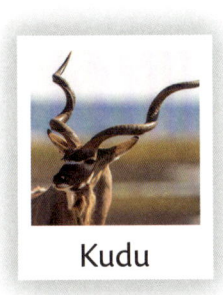

Babirusa

245

GOD MADE EVERYTHING

Hooves are attached to an animal's toes. Let's see if you can figure out why most hoofed mammals are fast runners.

1. Go outside where you have room to run fast.

2. Run like you usually do and pay attention to your feet. Did you use your toes to make yourself go faster?

   _____

3. Now try to run on your heels. Keep your toes off the ground. Could you run as quickly like this?

   _____

ACTIVITY 88: SPECIAL HEADS AND FEET

4. Why does toe-running help an animal run faster?

_____

_____

- Sing the hymn for this unit.

- Repeat the memory verse for this unit.

- Praise God for how beautiful horses look when they run!

UNIT 8 • CHAPTER 30

ACTIVITY 89

# Thirsty Camels

There was once a kind young woman named Rebekah. When Rebekah lived, people didn't have water coming out of faucets like we do now. In the evenings, Rebekah would walk to a well to get water.

One evening at the well, a thirsty traveler asked Rebekah for a drink. He was an old man and very tired. She happily gave him water from her pitcher. Then she offered to get water for his ten thirsty camels too! How kind she was to think of doing that without being asked.

Rebekah's kindness showed the traveler that God had led him to the right place. You can learn more about Rebekah from the Bible in Genesis 24.

Let's imagine how much water a thirsty camel can drink. Fill a container with one gallon (4 L) of water. Can you lift it? A thirsty camel can drink 40 gallons of water!

The camel on the next page is standing next to one gallon of water. On the camel coloring page, draw 40 containers of water around the camel. Color the picture.

## ACTIVITY 89: THIRSTY CAMELS

- Think about how much water Rebekah must have carried for ten camels!

- Sing the hymn for this unit.

- Repeat the memory verse for this unit.

- Praise God for Rebekah's kindness. Ask Him to show you ways you can be kind to people.

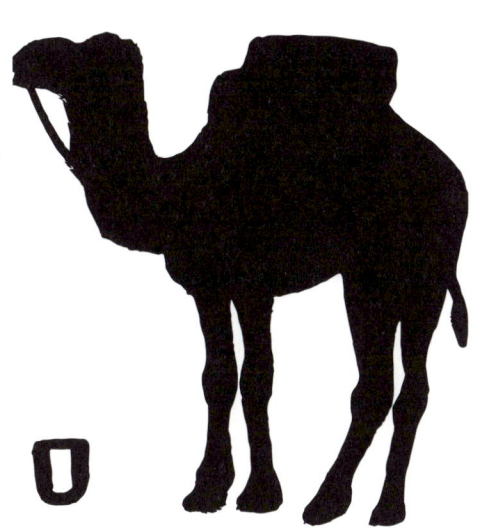

# GOD MADE EVERYTHING

ACTIVITY 89: THIRSTY CAMELS

UNIT 8 • CHAPTER 30
# ACTIVITY 90
# Color the Wild Hoofed Mammals

Two of these mammals have antlers. Color their antlers gray. Two of them have horns. Color their horns brown. Color the mammals' hair any color you like.

GOD MADE EVERYTHING

Sing the hymn for this unit.

Repeat the memory verse for this unit.

Praise God for so many different kinds of wild animals with hooves! Tell Him what your favorite kind of hoofed animal is.

UNIT 8 • CHAPTER 31

ACTIVITY 91
# Paw Pads

You have probably seen the paw pads on a pet dog or cat. Have you ever squeezed these pads and noticed their tough squishiness? This squishiness helps protect the animal's feet.

Let's crawl around to learn something about paw pads!

A dog's paw pads

1. Find a hard floor with no rugs or carpet on it. Crawl on your hands and knees on the floor. What gets sore sooner: your hands or your knees?

   _____

2. Look at one of your hands. Do you see the squishy muscle at the bottom of your thumb? There is another muscle on the other side of your hand down from your smallest finger. These muscles helped pad your hands as you were crawling. What's inside animal paw pads to make them squishy?

   _____

GOD MADE EVERYTHING

3. Now look at your knees where they hit the floor as you were crawling. Do you have any muscles there to pad your knees?

_____

4. Try crawling on your knuckles and knees. Which would you rather use to crawl: your hands or your knuckles?

_____ Why?

_____

_____

Sing the hymn for this unit.

Repeat the memory verse for this unit.

Thank God that He made walking comfortable for dogs, cats, and you!

256

UNIT 8 • CHAPTER 31

# ACTIVITY 92
# Hyenas Laugh and Eat

Meat eaters have to work hard for their food. They have to find, chase, catch, and sometimes fight their food before they can eat it. Each meal, God is the one who makes sure they are fed.

**The young lions roar after their prey
And seek their food from God.
(Psalm 104:21)**

Hyenas are interesting meat eaters. They are not cats. They are not dogs. But hyenas look and act a little like cats and dogs. They are sometimes called laughing hyenas because of the loud call they make when they invite their friends to a feast. Sometimes it's a feast they have killed. Sometimes it's leftovers from another animal's feast.

God gave hyenas the strongest jaws on earth! They need strong jaws because they use their mouths to crush and eat bones. After a group of lions eats all it wants from a hunt, the hyenas come and eat everything else. They don't mind if it's rotten. After the hyena's stomach gets all it needs from a meal, the hyena throws up the hooves and hair but not the bones.

Complete the dot-to-dot to see what a hyena looks like! Color the picture.

# GOD MADE EVERYTHING

Sing the hymn for this unit.

Repeat the memory verse for this unit.

Thank God that He gave hyenas the jaws they need for eating. Praise Him that hyenas help clean up the land!

ACTIVITY 92: HYENAS LAUGH AND EAT

259

# GOD MADE EVERYTHING

UNIT 8 • CHAPTER 31

# ACTIVITY 93
# Dry Some Fruit!

Pika looking for a sunny rock to dry its food on

God says it's wise to gather food. Just like the pika, people can gather food and store it to eat later.

**He who gathers in summer is a wise son;**
**He who sleeps in harvest is a son who causes shame.**
**(Proverbs 10:5)**

Let's dry some food! When food dries, little pieces of water go out of the food into the air. The food gets small and wrinkly. Without water, germs cannot grow in the food. When there are no germs in food, it doesn't get rotten. This means that dried food will last longer than normal food!

## Fruit Drying

**You will need:**

- An oven
- A baking sheet
- Baking parchment paper
- An adult with a knife
- Some fresh fruit that is ripe but not bruised or overripe. Some fruits that dry well are apples, pears, peaches, apricots, bananas, cherries, and berries

**Instructions:**

1. Wash the fruit.
2. Cherries and berries can be dried whole. The other fruits should be peeled and their cores or pits removed. Ask the adult to slice the fruit into pieces of the same thickness so they all dry at the same time.
3. Arrange the fruit pieces on the baking sheet so they're not touching each other.
4. Bake at 170 degrees F (77 C). Make sure the baking sheet is not touching the sides of your oven so air can move around it.
5. Keep the oven door cracked open, and stir the fruit every 30 minutes.
6. Dry for 4-8 hours or until the fruit is chewy. It should not be squishy or crispy.
7. Remove the baking sheet from the oven. Let the fruit sit at least 12 hours. Pack in storage containers to enjoy later.

ACTIVITY 93: DRY SOME FRUIT!

Sing the hymn for this unit.

Repeat the memory verse for this unit.

Thank God for providing a way to keep food from rotting so it can be eaten later!

UNIT 8 • CHAPTER 32

ACTIVITY 94

# Do the Elephant Walk and the Monkey Swing!

Baby Elephant Walk:

1. Ask an adult to search for the music "'Baby Elephant Walk' by Henry Mancini" on the internet. This is a fun piece of music to pretend with!

2. While the music is playing, pretend you are a baby elephant. Bend forward at your waist. To make your trunk, fold your hands together and let them hang straight down.

3. Walk heavily and swing your trunk in time to the music.

Monkey Swing

4. Find a playground bar or tree branch that you can wrap your hands around. Ask an adult to help with this activity.

5. Monkeys have thumbs like you do. Try hanging from the bar with your thumbs around the bottom of the bar and your fingers around the top. Can you swing back and forth

ACTIVITY 94: DO THE ELEPHANT WALK AND THE MONKEY SWING!

and side to side? _____ Can you let go with one hand and put it back again? _____

6. Now try to hang on without using your thumbs. It's a lot harder to do this, isn't it? _____

7. Do you see why God gave monkeys, apes, and lemurs special thumbs and toes? _____

- Sing the hymn for this unit.

- Repeat the memory verse for this unit.

- Thank God for elephants and for making special thumbs for monkeys and for us!

UNIT 8 • CHAPTER 32

**ACTIVITY 95**

# Sea Mammals

Color the sea mammals. Circle the one that is not a whale.

Sing the hymn for this unit.

Repeat the memory verse for this unit.

Praise God for all the interesting sea mammals He made! Thank Him for your favorite one.

ACTIVITY 95: SEA MAMMALS

GOD MADE EVERYTHING

UNIT 8 • CHAPTER 32
## ACTIVITY 96
# Hang Like a Bat!

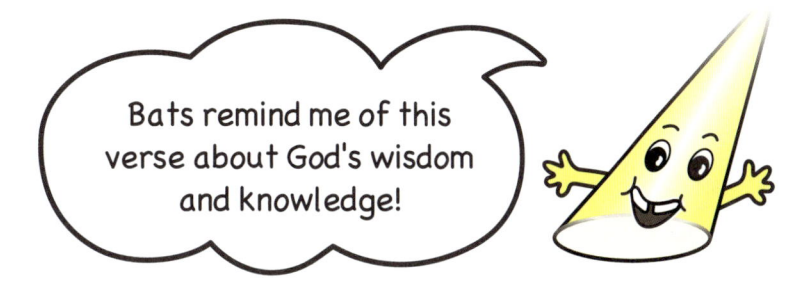

*Bats remind me of this verse about God's wisdom and knowledge!*

**The darkness and the light are both alike to You. (Psalm 139:13)**

Bats hang upside down to sleep. Let's go outside and pretend to be a sleeping bat! Take along an adult to help.

1. Find the playground bar or tree branch you used when you pretended to be a monkey.

2. Grab the bar with your hands. Ask an adult to help you bring your feet through your arms and over the bar so you can hang from your knees.

3. Once you are upside down, fold your arms around yourself like bat wings.

4. Close your eyes to make it seem dark like a cave.

- Sing the hymn for this unit.

- Repeat the memory verse for this unit.

- Thank God that bats can sleep upside down. Thank Him that you get to sleep flat in a bed.

UNIT 9 • CHAPTER 33
ACTIVITY 97

# Made in His Image

**So God created man in His own image. (Genesis 1:27)**

When you look in a mirror, you are seeing your **image**. Your reflection looks like you, but it is not you. God made people in His image. He made us to be like Him in some ways.

One way people are special is that we can make things. Let's create a picture about Psalm 104. Make a collage with a collection of nature pictures. Ask an adult to help you gather pictures from magazines or the internet.

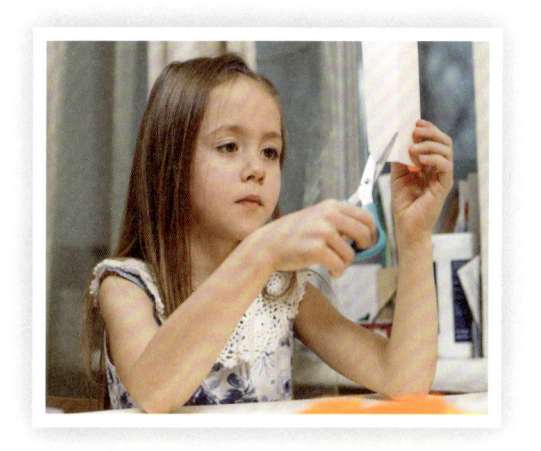

1. Ask someone to read Psalm 104 to you. This psalm talks about many of the things God created.

2. Make your creation collage by cutting out nature pictures and gluing them to a paper in a nice design.

GOD MADE EVERYTHING

Sing the hymn for this unit.

Write or repeat the memory verse for this unit.

### Memory Verse

I will praise You, for I am fearfully and wonderfully made.
(Psalm 139:14)

Praise God that He blessed us by making us in His image!

UNIT 9 • CHAPTER 33
ACTIVITY 98

# There's No One Else Like You!

Read Psalm 139:1-6 to see how special you are to God.

Use an ink pad to make a set of your own special fingerprints!

1. Trace one of your hands inside the square outline.

2. Press the thumb of the same hand onto the ink pad. Then press your thumb onto the traced thumb on the paper below. Can you see your thumbprint on the paper?

3. Make prints of each of your fingers in the same way.

Sing the hymn for this unit.

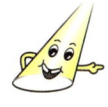

Repeat the memory verse for this unit.

Praise God for making each person different!

# GOD MADE EVERYTHING

274

## ACTIVITY 98: THERE'S NO ONE ELSE LIKE YOU!

# GOD MADE EVERYTHING

UNIT 9 • CHAPTER 33
ACTIVITY 99

# You Were Once a Newborn

**Before I formed you in the womb I knew you. (Jeremiah 1:5)**

- Color the picture of the mother and her newborn baby.

- Sing the hymn for this unit.

- Repeat the memory verse for this unit.

- Thank God for wonderfully making you from a tiny cell into a newborn baby. Thank Him for making you grow each day!

# GOD MADE EVERYTHING

# ACTIVITY 99: YOU WERE ONCE A NEWBORN

# GOD MADE EVERYTHING

UNIT 9 • CHAPTER 34

ACTIVITY 100
# Neuron Chain Tag

Messages go to and from your brain through a chain of neurons. This chain helps your body work better. If there was just one long neuron sending a message, its electricity would get tired and slow down. A neuron chain lets each neuron pass along the message to the next one before it gets tired.

Let's play Neuron Chain Tag. For this game, you will need at least three players, but you can use as many players as you want. Pick one player to be the first neuron. This player must try to tag another player. A tagged player must hold the hand of the first player, and together they have to chase the other players. The newest neuron to join the chain must be the one to tag the next neuron. As more and more players are tagged, they are added to the chain of neurons. The game ends when all the players are part of the chain. The last player to be tagged becomes the brain.

On the next page, circle the picture that shows the boy's brain and nerves.

GOD MADE EVERYTHING

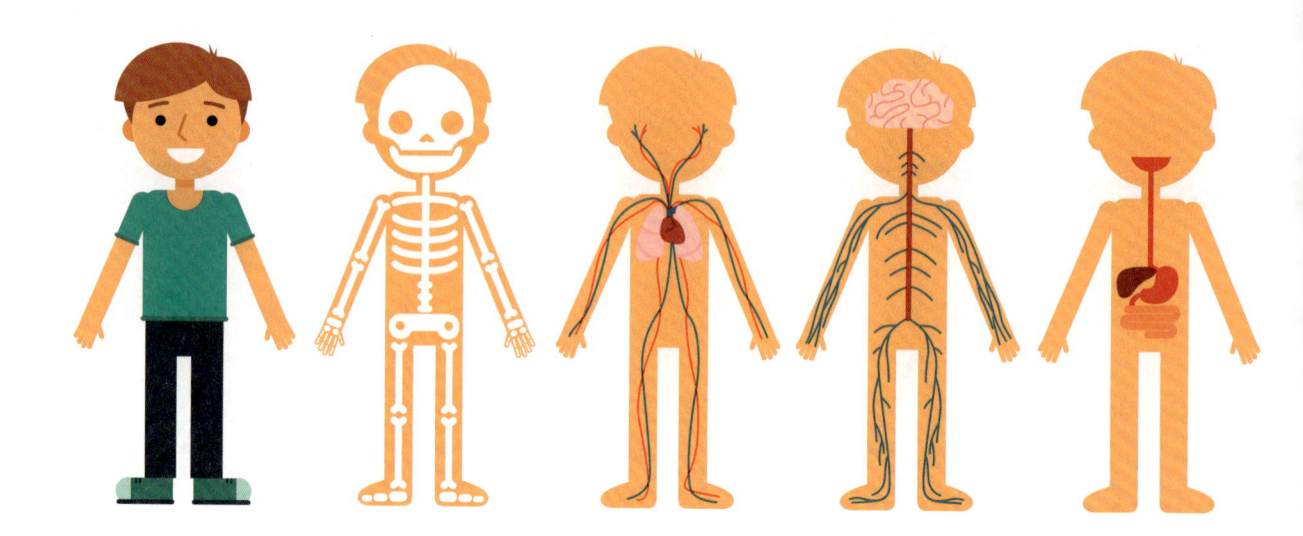

- Wear a helmet when you ride a bike, skateboard, or when you roller skate. Helmets protect your brain when you fall.

- Sing the hymn for this unit.

- Repeat the memory verse for this unit.

- Praise God for all the messages speeding around your body and brain through chains of neurons!

UNIT 9 • CHAPTER 34
## ACTIVITY 101
# Sound Tag

**The hearing ear and the seeing eye,
The LORD has made them both.
(Proverbs 20:12)**

Sound tag is a game that will show you how your ears can help find where a sound is coming from.

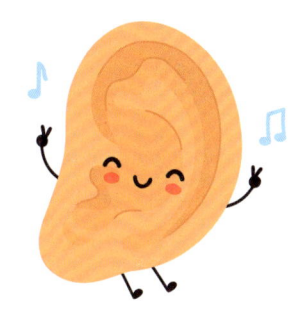

For Sound Tag, you will need at least three players. One person is "It." The other players must stay within a certain space (like a small backyard). The object of the game is for "It" to tag any other player. But "It" must keep his or her eyes closed. To help find the other players, "It" shouts, "Ear." The other players must shout back the word "Drum." Using these sounds, "It" tries to tag someone. A person who is tagged becomes the new "It."

Protect Your Hearing!

- Don't put things in your ears. You may damage your eardrum or push back earwax that is trying to come out.

- Don't listen to loud sounds or loud music. You may wear out the inside of your ear and lose some hearing when you get older.

GOD MADE EVERYTHING

Sing the hymn for this unit.

Repeat the memory verse for this unit.

Think of some sounds you like to hear. Thank God for these sounds!

UNIT 9 • CHAPTER 34
ACTIVITY 102
# The Tongue's Jobs

Your tongue has other jobs besides tasting!

Your tongue helps you eat. Get a snack and try to chew it without moving your tongue.

- Do you see how your tongue moves food around in your mouth so that your teeth can chew it? _____

- Can you swallow without using your tongue? _____

- Did you notice that the back of your tongue must move for you to swallow your food? _____

Your tongue helps you talk and sing. Pay attention to the way your tongue moves while you sing the hymn and repeat the memory verse for this unit.

To talk and sing, you must also make noise with your **voice box**. Your voice box is a part of your throat. You can make it

285

# GOD MADE EVERYTHING

vibrate as air goes through it. The vibration makes a sound—your voice! If you talk without making your voice box vibrate, you are whispering. Feel your throat as you talk, sing, and whisper. Voice boxes are wonderful! Our words and voices are very important to God. He loves to hear us sing to Him!

**Sing to God, sing praises to His name;**
**Extol Him who rides on the clouds.**
**(Psalm 68:4)**

Sing the hymn for this unit.

Repeat the memory verse for this unit.

Use your tongue and voice to praise God for creating you to talk and sing!

UNIT 9 • CHAPTER 35

# ACTIVITY 103
# Growing

How can you tell if you are growing? Does it seem like your clothes are getting shorter and tighter? Can you reach things that used to be out of reach? Are you strong enough now to open a door that used to be too heavy? These are ways you can tell that you are growing. Another way is to be measured once a year and see how you change.

Gather a measuring tape, a bathroom scale, and a helper. Ask your helper to measure you these ways:

- Height _____

- Weight _____

- Distance around your head _____

- Distance between your thumb and smallest finger when you stretch your hand _____

- Distance you can jump from a standstill _____

GOD MADE EVERYTHING

Sing the hymn for this unit.

Repeat the memory verse for this unit.

Thank God that you are growing and able to do more than when you were small. Ask Him to help you grow like Samuel did too!

UNIT 9 • CHAPTER 35

ACTIVITY 104
# Take a Breath

God makes you breathe automatically, but you can also breathe on purpose. When someone says, "Take a breath," you can purposely take a breath. Now let's pay attention to your breathing.

- Sit down and sit still for two minutes. Then look at your tummy area. It should be moving up and down as you breathe. You have a muscle under your lungs. This muscle pulls your lungs down. As it does, it makes your lungs fill with air. Even though your lungs are in your chest, they are being pulled down into your tummy and making your tummy go up and down. You are probably breathing only through your nose while you rest.

- Now run around outside until you are very tired. Sit down in the same seat and watch your breathing. Now your chest is also moving in and out with each breath. You are probably breathing through both your mouth and nose. Why are your tummy and chest both moving? Your body is trying to get more air because your muscles used up a lot of oxygen when you ran.

- On the next page, circle the picture that shows the parts of the body used in the blood cycle.

GOD MADE EVERYTHING

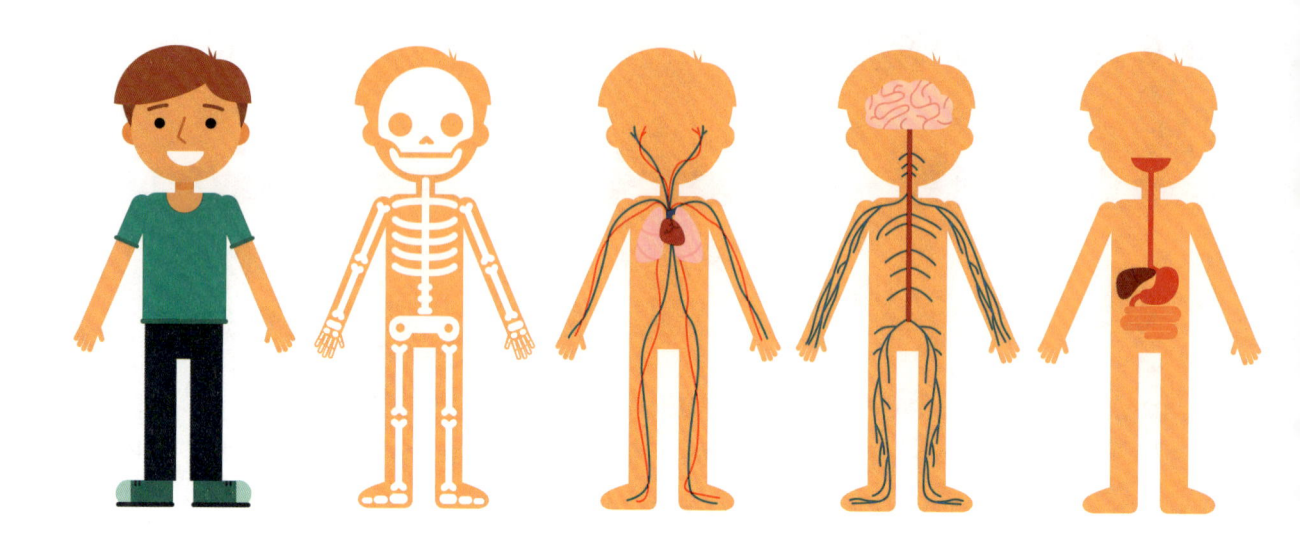

Sing the hymn for this unit.

Repeat the memory verse for this unit.

Thank God that you don't have to think about breathing. Praise Him that you breathe the right amount of air for the oxygen your body needs!

UNIT 9 • CHAPTER 35

# ACTIVITY 105
# Listen to Your Stomach

> "[He fills] our hearts with food and gladness." (Acts 14:17)

Sometimes people rest after eating a meal. But that's when the stomach gets busy!

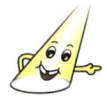

- Let's see if you can hear your stomach doing its job. You will need a glass of water and a roll of paper towels to listen through. You will hear better if the roll still has a lot of towels on it to block outside noise. Take your water and paper towels to a quiet room where there isn't any noise. Get a big mouthful of water but don't swallow it yet. Put one end of the cardboard tube on the middle of your stomach just under your ribs. Put the other end to your ear. You will have to turn your head. Now swallow the water and wait a few seconds. Did you hear when your stomach opened up and let the water in? What did it sound like? Did you hear gurgling? _____
You can try listening again after you eat a meal to hear your stomach working.

- On the next page, circle the picture that shows the parts of the body you use to digest food.

- Sing the hymn for this unit.

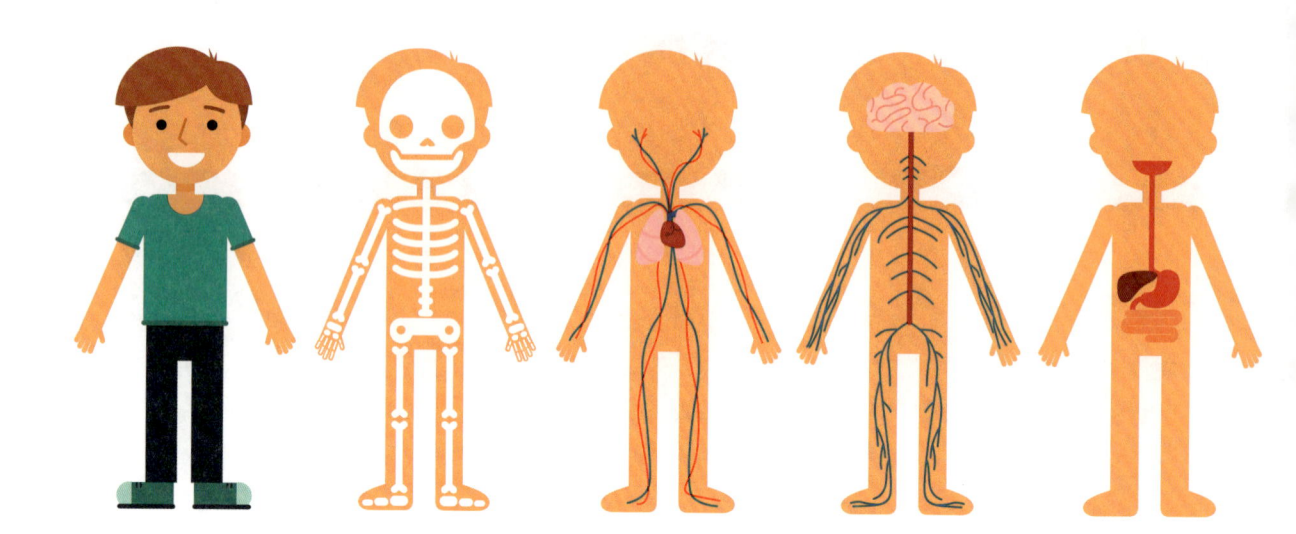

GOD MADE EVERYTHING

Repeat the memory verse for this unit.

Praise God for food, eating, and digestion! Thank Him for your favorite food.

UNIT 9 • CHAPTER 36
## ACTIVITY 106
# Your Skeleton

Your **skeleton** is what we call all of your bones connected together. God connected your bones together at **joints**. Joints are where your body bends. For these activities, you will need: a yardstick or broomstick, string, some paper, and 10-20 white pipe cleaners.

Let's see what it would be like if you had no joint at one of your knees. Using the string, tie a yardstick or broomstick to one of your legs above and below your knee to keep your leg from bending. Ask an adult if you need help. Try to walk, sit, and get up from a chair. Is it difficult to move without using the joint at your knee? _____

Circle the picture of the boy's skeleton.

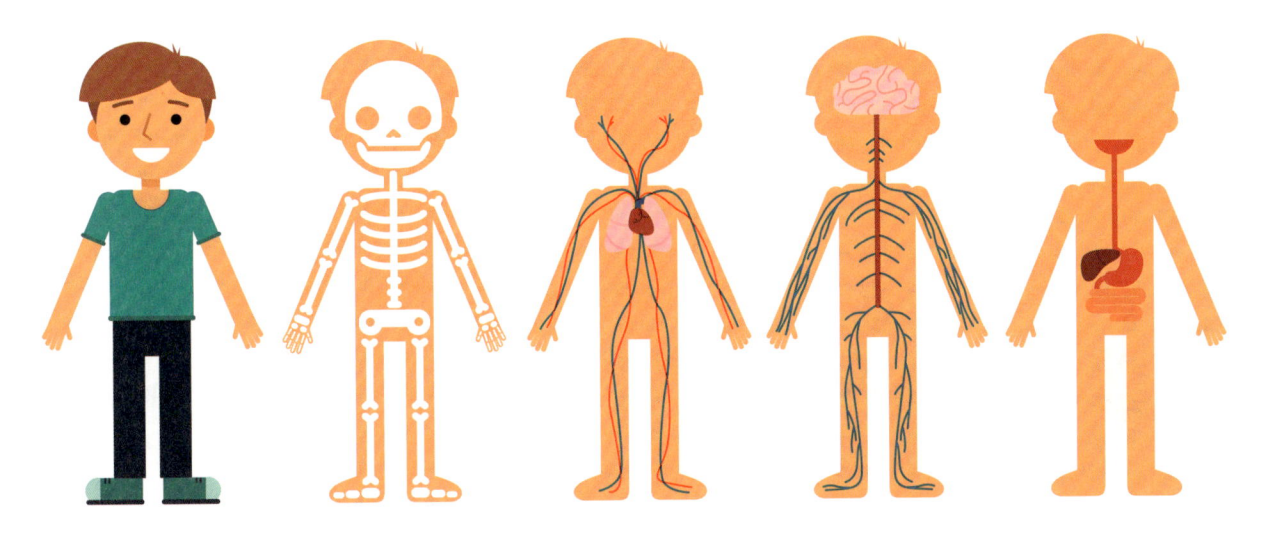

293

GOD MADE EVERYTHING

Make a skeleton by twisting pipe cleaners together. For the skull, wind pipe cleaners around some crumpled paper. Try to only make the skeleton bent where joints would be.

Sing the hymn for this unit.

Repeat the memory verse for this unit.

Thank God for your own sturdy skeleton and the joints that let it bend!

**To an adult helper**

Please see Activity 108 for instructions on preparing ahead of time.

UNIT 9 • CHAPTER 36
## ACTIVITY 107
# Using Muscles

You use your muscles all day long. Muscles are what make you move. When you run, you are using many of your muscles. When you sit still, you are using muscles to hold yourself up. You have 42 muscles in your face to smile or cry with.

"For in Him we live and move and have our being." (Acts 17:28)

Sit on a small chair or step with your feet on the floor. Put your hands on the tops of your thighs so you can feel your leg muscles. What happens to these muscles when you try to stand up?

*God is the one who gives you life and makes you move!*

GOD MADE EVERYTHING

Put a circle around the girl that is using her muscles the most. Put a square around the girl that is using her muscles the least.

Sing the hymn for this unit.

Repeat the memory verse for this unit.

Praise God that He gave you strong muscles to play, work, and smile with.

296

UNIT 9 • CHAPTER 36

# ACTIVITY 108
# Trying to Learn

Learning does not happen automatically. It takes work to learn. You have probably been working on your memory verse by repeating it after it is read to you. Have you noticed that you don't learn the verse as easily if you are thinking of something else while you are repeating it? Today's activity will help you see how important it is to try when you are learning.

> **Secret instructions for an adult helper**
>
> 1. Without drawing attention to what you are doing, put a selection of commonplace objects where your student will see them but not have an opportunity to study them. This could be a place he passes by or a place where he will be busy doing something else.
> 2. When you know he has had an opportunity to notice the objects, cover them and ask him to name as many as he can remember. How many items were remembered? _____
> 3. Uncover the items and tell him he is allowed one minute to look at them and that you will be asking him to name as many as he can afterward.
> 4. After the minute, cover the items and ask your student to name the objects he can remember. How many items were remembered? _____

GOD MADE EVERYTHING

**Whatever your hand finds to do, do it with your might. (Ecclesiastes 9:10)**

Answer these questions!

1. Did you remember more items after you were told that you would be asked to name them? _____

ACTIVITY 108: TRYING TO LEARN

2. If so, how many more did you remember the second time you were asked? _____

3. Do you see that you can learn better if you try to learn? _____

Sing the hymn for this unit.

Repeat the memory verse for this unit.

Praise God for all you have learned about His creation! Thank Him for giving us the Bible so we can learn His wisdom.

# ANSWER KEY

**ACTIVITY 2.** Two boats

**ACTIVITY 3.** You should see flashes of white when the spinner spins its fastest. White light is made of all the colors. You may see a tint of color depending on the purity, brightness, and balance of colors used, but it should be very light.

**ACTIVITY 4.** Since Beamer is light, he can't block light to make a shadow., Circle picture of girl on camel.

**ACTIVITY 8.** 1. Plants, 2. plants, 3. plants

**ACTIVITY 9.** 1. O, 2. M, 3. M, 4. M, 5. O, 6. M, 7. O

**ACTIVITY 10.** 1. Dark, 2. Damp, 3. No

**ACTIVITY 11.** 1. God, 2. Me, 3. Me, 4. God, 5. Me, 6. God, 7. God, 8. Me, 9. God, 10. God

**ACTIVITY 12.** 1. My pupils opened (got bigger) to let in more light., 2. My pupils closed (got smaller)., 3. To protect my eyes from too much light, 4. Horse, 5. Cat, 6. Dog

**ACTIVITY 15.** Gravity

**ACTIVITY 16.** Wise: House on rock, Foolish: House on sand, Circle the wise man, Dozen=12

**ACTIVITY 20.** Eyes

**ACTIVITY 25.** 1. Yes, No, 2. No, Yes, 3. Yes (light usually puts out heat), Yes, 4. No, No, 5. No, Yes

**ACTIVITY 34.** 1. Comet, 2. Meteoroid, 3. Meteor, 4. Meteorite

**ACTIVITY 44.** Oxygen on top, Carbon dioxide on bottom

**ACTIVITY 45.**

| Y | G | R | G |
| B | G | B | R |
| O | Y | Y | O |
| R | O | R | B |

**ACTIVITY 52.** 1. 3, 2. 2, 3. 6, 4. 4

**ACTIVITY 53.** Arrows go clockwise, 1. Moth and butterfly, 2. Moth, 3. Grasshopper

**ACTIVITY 55.** Green

**ACTIVITY 57.** 1. B, 2. D, 3. M, 4. F, 5. T, 6. W, 7. G, 8. P

**ACTIVITY 61.** 1. Empty, 2. Full

**ACTIVITY 67.** Crocodile rhymes with reptile.

# GOD MADE EVERYTHING

**ACTIVITY 68.**

**ACTIVITY 69.** The picture on bottom left is not a tortoise. Since it has flippers, it's a turtle.

**ACTIVITY 75.** The path of food starts at the beak and travels through the yellow-colored tubes and organs. 1. Gizzard, 2. Stomach, 3. Intestine, 4. Crop

**ACTIVITY 77.** #3

**ACTIVITY 81.**

**ACTIVITY 82.** The water runs easily off the oily hand.

**ACTIVITY 84.** God's Words are sweeter than honey.

**ACTIVITY 87.** 1. Joey, 2. Puppy, 3. Foal, 4. Pup, 5. Calf, 6. Kit, 7. Kitten, 8. Cub, 9. Fawn

**ACTIVITY 88.** 1. Babirusa, 2. Kudu, 3. Reindeer; Toe-running gives an animal extra spring to take long steps.

**ACTIVITY 90.** Antlers on left, horns on right

**ACTIVITY 91.** 1. Probably "knees," 2. Fat, 3. Probably "no," 4. Hands, More padding on hands

**ACTIVITY 94.** All answers probably "yes"

**ACTIVITY 100.** Fourth from the left

**ACTIVITY 102.** Probably "yes," "no," "yes"

**ACTIVITY 104.** Middle picture

**ACTIVITY 105.** Last picture on right

**ACTIVITY 106.** Second from left

**ACTIVITY 107.** The muscles get hard, Playing tennis (circle) uses the most muscles, sleeping (square) use the least.